VOLUME ONE HUNDRED AND SEVENTEEN

INTERNATIONAL REVIEW OF NEUROBIOLOGY

Cerebellar Conditioning and Learning

INTERNATIONAL REVIEW OF NEUROBIOLOGY

VOLUME 117

SERIES EDITORS

R. ADRON HARRIS

Waggoner Center for Alcohol and Drug Addiction Research
The University of Texas at Austin
Austin, Texas, USA

PETER JENNER

Division of Pharmacology and Therapeutics
GKT School of Biomedical Sciences
King's College, London, UK

VOLUME ONE HUNDRED AND SEVENTEEN

INTERNATIONAL REVIEW OF NEUROBIOLOGY

Cerebellar Conditioning and Learning

Edited by

MICHAEL D. MAUK

Center for Learning and Memory, and
Department of Neuroscience,
The University of Texas,
Austin, Texas, USA

ELSEVIER

AMSTERDAM • BOSTON • HEIDELBERG • LONDON
NEW YORK • OXFORD • PARIS • SAN DIEGO
SAN FRANCISCO • SINGAPORE • SYDNEY • TOKYO
Academic Press is an imprint of Elsevier

AP

Academic Press is an imprint of Elsevier
32 Jamestown Road, London NW1 7BY, UK
525 B Street, Suite 1800, San Diego, CA 92101-4495, USA
225 Wyman Street, Waltham, MA 02451, USA
The Boulevard, Langford Lane, Kidlington, Oxford OX5 1GB, UK

First edition 2014

Notices

Knowledge and best practice in this field are constantly changing. As new research and experience broaden our understanding, changes in research methods, professional practices, or medical treatment may become necessary.

Practitioners and researchers must always rely on their own experience and knowledge in evaluating and using any information, methods, compounds, or experiments described herein. In using such information or methods they should be mindful of their own safety and the safety of others, including parties for whom they have a professional responsibility.

To the fullest extent of the law, neither the Publisher nor the authors, contributors, or editors, assume any liability for any injury and/or damage to persons or property as a matter of products liability, negligence or otherwise, or from any use or operation of any methods, products, instructions, or ideas contained in the material herein.

ISBN: 978-0-12-420247-4
ISSN: 0074-7742

For information on all Academic Press publications
visit our website at store.elsevier.com

CONTENTS

CONTRIBUTORS

John H. Freeman
Department of Psychology and Neuroscience Program, The University of Iowa, Iowa City, Iowa, USA

Giorgio Grasselli
Department of Neurobiology, University of Chicago, Chicago, Illinois, USA

Hunter Halverson
Center for Learning and Memory, The University of Texas, Austin, Texas, USA

Christian Hansel
Department of Neurobiology, University of Chicago, Chicago, Illinois, USA

Andrei Khilkevich
Center for Learning and Memory, The University of Texas, Austin, Texas, USA

Wenke Li
Center for Learning and Memory, The University of Texas, Austin, Texas, USA

Michael D. Mauk
Center for Learning and Memory, and Department of Neuroscience, The University of Texas, Austin, Texas, USA

Hiroshi Nishiyama
Center for Learning and Memory, Department of Neuroscience, University of Texas at Austin, Austin, Texas, USA

CHAPTER ONE

Learning-Induced Structural Plasticity in the Cerebellum

Hiroshi Nishiyama[1]
Center for Learning and Memory, Department of Neuroscience, University of Texas at Austin, Austin, Texas, USA
[1]Corresponding author: e-mail address: hiroshi@utexas.edu

Contents

Abstract

Activity-dependent changes in synaptic properties are considered key neural mechanisms of learning and memory. Most studies focus on changes in synaptic function such as long-term potentiation (LTP) and long-term depression (LTD), while changes in synaptic structure have been largely ignored. However, structural synaptic changes are also important. In fact, LTP and LTD are often associated with structural alterations of dendritic spines. Furthermore, experimental evidence indicates that behavioral learning often induces structural rewiring of synaptic circuitry through the formation of new synapses and elimination of existing ones. To discuss the role of structural plasticity in cerebellar learning, this chapter mainly focuses on parallel fiber–Purkinje cell synapses in the cerebellar cortex and reviews their structural changes under several different forms of motor learning.

International Review of Neurobiology, Volume 117
ISSN 0074-7742
http://dx.doi.org/10.1016/B978-0-12-420247-4.00001-4

1. INTRODUCTION

Can experience change brain structures? Fifty years ago, Bennett, Diamond, Krech, and Rosenzweig (1964) addressed this century-old question with carefully designed animal experiments. They housed littermates of rats (same sex and age) in two different conditions, enriched environment versus restricted environment, and found a highly consistent and reproducible increase in the cortical weight in the enriched environment group (Bennett et al., 1964). This finding, along with other observations, allowed them to speculate that synaptic connections were substantially increased by experience. After this initial discovery, a number of studies convincingly showed that various aspects of experience and biological rhythms can alter synaptic connectivity through structural rearrangement of neurons in the adult brain (Black, Isaacs, Anderson, Alcantara, & Greenough, 1990; Chang & Greenough, 1982; Geinisman, Berry, Disterhoft, Power, & Van der Zee, 2001; Greenough, Larson, & Withers, 1985; Jones, Klintsova, Kilman, Sirevaag, & Greenough, 1997; Knott, Quairiaux, Genoud, & Welker, 2002; Popov & Bocharova, 1992; Popov, Bocharova, & Bragin, 1992; Woolley, Gould, Frankfurt, & McEwen, 1990).

Despite these findings, the roles of neuronal structural plasticity in learning and memory have been largely unclear. First, behavioral treatments used in earlier studies involved a large amount of nonlearning components, such as difference in social interaction, substantial repetition of movements, and change in overall locomotion activity (Bennett et al., 1964; Chang & Greenough, 1982; Greenough et al., 1985). Second, in most previous studies, a between-subject comparison was used to infer what occurred in the brain during given behavioral tasks. In other words, the observed differences between control and experimental groups did not directly represent how the same neurons sequentially changed their structure over time. Because of these shortcomings, our understandings of neuronal structural plasticity in learning advanced, until recently, at a relatively slow pace. Meanwhile, the discovery of hippocampal long-term potentiation (LTP) in 1973 opened the era of functional plasticity research (Bliss & Lomo, 1973). The roles of LTP and long-term depression (LTD) in learning and memory have been intensively studied ever since. In most of those studies, long-term functional plasticity of synapses is the central theme, whereas structural plasticity received less attention.

This situation has changed during the last decade. Owing largely to advances in high-resolution live imaging and three-dimensional reconstruction of serial electron micrographs, it has been clearly shown that LTP and LTD are accompanied by structural plasticity of dendritic spines in the hippocampus (Engert & Bonhoeffer, 1999; Matsuzaki, Honkura, Ellis-Davies, & Kasai, 2004; Nagerl, Eberhorn, Cambridge, & Bonhoeffer, 2004; Okamoto, Nagai, Miyawaki, & Hayashi, 2004; Ostroff, Fiala, Allwardt, & Harris, 2002; Zhou, Homma, & Poo, 2004). Furthermore, the advent of two-photon *in vivo* time-lapse microscopy enabled imaging of the same neurons repeatedly over a long period of time in the brains of live animals (Grutzendler, Kasthuri, & Gan, 2002; Lendvai, Stern, Chen, & Svoboda, 2000; Trachtenberg et al., 2002). Such *in vivo* time-lapse imaging experiments have revealed that (i) a small fraction of presynaptic axonal boutons, dendritic spines, and even dendritic arbors remain dynamic in adulthood even without any sensory or behavioral manipulation (De Paola et al., 2006; Grutzendler et al., 2002; Holtmaat et al., 2005; Lee et al., 2006; Lendvai et al., 2000; Majewska, Newton, & Sur, 2006; Nishiyama, Fukaya, Watanabe, & Linden, 2007; Stettler, Yamahachi, Li, Denk, & Gilbert, 2006; Trachtenberg et al., 2002; Zuo, Lin, Chang, & Gan, 2005) and (ii) novel sensory experience and motor learning promote the formation and elimination of dendritic spines in the adult neocortex (Holtmaat, Wilbrecht, Knott, Welker, & Svoboda, 2006; Xu et al., 2009; Yang, Pan, & Gan, 2009; Zuo, Yang, Kwon, & Gan, 2005). All these technical advances provide us with tools to examine the roles of structural plasticity in learning and memory with unprecedented levels of details.

Cellular mechanisms of learning and memory have been extensively studied in the cerebellum. Different cerebellar neurons show a wide variety of functional plasticities, including LTP and LTD, and changes in intrinsic excitability (Carey, 2011; Gao, van Beugen, & De Zeeuw, 2012; Hansel, Linden, & D'Angelo, 2001). However, the prominence of functional plasticity research appears to overshadow the contribution of structural plasticity to learning in the cerebellum. This chapter therefore focuses on structural plasticity and discusses its potential roles in learning.

2. LEARNING-INDUCED STRUCTURAL PLASTICITY AT PARALLEL FIBER–PURKINJE CELL SYNAPSES

Purkinje cells are inhibitory projection neurons in the cerebellar cortex. They provide the sole cortical output to the deep cerebellar nuclei that send motor commands to various motor centers located outside of the cerebellum. Therefore, the activity of Purkinje cells is considered the final outcome of neural computation in the cerebellar cortex, which eventually influences motor behavior of animals (Eccles, Ito, & Szentagothai, 1967). Each Purkinje cell receives excitatory synaptic inputs from two different sources, climbing fibers and parallel fibers, and the parallel fiber–Purkinje cell synapses show postsynaptically expressed, bidirectional (i.e., potentiation and depression), long-term synaptic plasticity (Jorntell & Hansel, 2006). LTD is induced when parallel fibers and climbing fibers are concurrently activated (Ito & Kano, 1982; Ito, Sakurai, & Tongroach, 1982), whereas LTP is induced when only parallel fibers are activated (Coesmans, Weber, De Zeeuw, & Hansel, 2004; Lev-Ram, Wong, Storm, & Tsien, 2002). Since the roles of LTD and LTP (mostly LTD) in motor learning are intensively investigated (Ito, 2001; Mauk, Garcia, Medina, & Steele, 1998), it is reasonable to ask whether (i) this synapse also shows structural plasticity and, if so, (ii) if it is associated with learning.

2.1. Acrobatic motor skill learning

As described above, a conventional experimental approach for studying experience-dependent structural plasticity relies upon analysis of fixed tissue, which necessitates between-animal comparisons. Since analysis of this type is not sensitive enough to detect subtle changes, behavioral tasks were often designed such that massive and widespread structural changes were expected. In such behavioral tasks, dissociating learning and nonlearning components of behavior are not always straightforward. But in 1990, Greenough and coworkers addressed this problem by devising an acrobatic motor skill learning task (Black et al., 1990). In this task, animals are required to sequentially traverse various elevated obstacles leading to their home cage. During the training over several weeks, animals show substantial improvement in their performance (measured by the latency to traverse individual obstacles), indicating an improvement in their sensory–motor integration and whole-body and limb coordination. Importantly, this acrobatic motor skill learning does not substantially increase overall motor activity. Furthermore, several different control tasks were carefully designed to dissociate motor learning from the increase in overall motor activity. Their data showed that acrobatic motor skill learning induced synaptogenesis in the cerebellar cortex; the number of synapses per Purkinje cell in the paramedian lobule was increased by approximately 25%. On the other hand, an increase of overall motor activity by repetition of same simple movements (i.e., motor learning was not involved) did not induce synaptogenesis. These results indicate that learning new motor skills can be mediated at least in part by structural rewiring of synaptic connections in the cerebellar molecular layer.

Besides the excitatory climbing fiber and parallel fiber, Purkinje cells receive inhibitory inputs from molecular layer interneurons. A subsequent study from the same group showed that the learning-induced increase in synapses selectively occurred at parallel fiber–Purkinje cell synapses (Kleim et al., 1998). Neither climbing fiber–Purkinje cell nor interneuron–Purkinje cell synapses were affected. This input-specific parallel fiber structural plasticity is interesting, because many previous studies suggest that LTD at this synapse plays an important role in simple forms of motor learning, such as adaptation of vestibule–ocular reflex and eyelid conditioning (Ito, 2001; Raymond, Lisberger, & Mauk, 1996). However, it is unclear (i) whether LTD is involved in acrobatic motor skill learning and (ii) how the increase in parallel fiber–Purkinje cell synapses contributes to the improvement of complex motor skills. It appears difficult to address the former question, because multiple different brain regions are most likely involved in acrobatic motor skill learning. In fact, acrobatic learning is known to induce structural plasticity in the motor cortex as well (Jones, Chu, Grande, & Gregory, 1999). Therefore, even if LTD at parallel fiber–Purkinje cell synapses plays a role in acrobatic learning in normal circumstances, blocking LTD does not necessarily lead to a deficit in learning because plasticity in other brain regions may compensate for the lack of LTD.

The latter question (how the increase of parallel fiber–Purkinje cell synapses contributes to learning) is also difficult to address, but some insights can be obtained by more detailed analyses of structural plasticity at parallel fiber–Purkinje cell synapses. The increase in the total number of parallel fiber–Purkinje cell synapses is primarily caused by the increase in volume of the molecular layer, not the density of the synapses (Black et al., 1990; Kleim et al., 1998). Since the density of spines is already very high on normal Purkinje cell dendrites, dendritic volume might need to expand for producing additional spines to accommodate to the 25% increase of the synapses. However, the lack of significant change in synapse density does not mean that preexisting synapses are unaffected. Most parallel fiber *en passant* boutons contact a single Purkinje cell spine, but acrobatic motor skill learning induces the formation of multisynaptic boutons that contact multiple spines (Federmeier, Kleim, & Greenough, 2002; Pichitpornchai, Rawson, & Rees, 1994). In addition, both the length and density of Purkinje cell spines are slightly (~10%) increased by acrobatic learning (Lee, Jung, Arii, Imoto, & Rhyu, 2007). Therefore, acrobatic learning is associated with both global and local changes in synaptic structures. Here, the global change indicates the increase in postsynaptic volume, while the local change indicates the conversion from monosynaptic to multisynaptic presynaptic boutons and the morphological alterations of postsynaptic spines.

Recently, studies by our group and others provided significant new insights into this local structural plasticity (Carrillo, Cheng, Ko, Jones, & Nishiyama, 2013; Lee et al., 2013). We used two-photon *in vivo* time-lapse microscopy to repeatedly image the same sets of parallel fiber boutons over several months while animals were performing acrobatic motor skill learning. We found that (i) in control conditions without learning, ~10% of presynaptic boutons were formed and eliminated over a period of 2 weeks and the formation and elimination of boutons were nearly balanced and (ii) acrobatic learning significantly reduced the formation of new boutons, but not the elimination of existing boutons (Carrillo et al., 2013). These results indicate that acrobatic learning is associated with the reduction of basal presynaptic structural plasticity and a slight decrease in the total number of boutons. We speculate that parallel fiber structural plasticity is reduced because available intracellular resources within the fibers are preferentially used to convert a selected group of existing monosynaptic boutons to multisynaptic boutons rather than to create new boutons. Furthermore, parallel fiber structural plasticity may contribute to acrobatic learning by strengthening selected inputs via the formation of multisynaptic boutons while maintaining the overall strength of synaptic drive at relatively constant levels by reducing the total number of boutons (Carrillo et al., 2013).

Another study, which was performed independently of ours, yielded a very similar overall conclusion. Lee et al. (2013) used three-dimensional reconstruction of serial electron micrographs to examine the dendritic origin of Purkinje cell spines that contact multisynaptic boutons. They found that (i) acrobatic learning selectively increased those multisynaptic boutons that contacted spines arising from the same dendritic segment and (ii) synapses near multisynaptic boutons were smaller in acrobatic learning animals compared with control animals (Lee et al., 2013). The former finding is consistent with the idea that the formation of multisynaptic boutons strengthens local synaptic efficacy, while the latter suggests compensatory weakening of adjacent synapses to maintain the constant level of overall synaptic activity within local dendritic segments. Thus, two independent studies, using different techniques (live imaging vs. electron microscopy) for different focuses (pre- vs. postsynaptic plasticity), similarly suggest that parallel fiber–Purkinje cell synapse is a highly tuned system that allows selective strengthening and weakening of synaptic efficacy through balanced structural plasticity. Such a system appears to be suitable for the fine-tuning of motor behavior that is required for acrobatic motor skill learning.

2.2. Eyelid conditioning

Eyelid conditioning is a form of classical conditioning in which the association between a neutral, conditioned stimulus (CS, such as a tone) and an unconditioned stimulus (US, such as an air puff to the eye) is learned. An eyelid response is always elicited by the US, whereas a CS does not elicit response in naive animals. But a repetitive presentation of a CS–US pair (the CS is always followed by the US) to animals eventually allows the CS to elicit precisely timed eyelid closure that occurs immediately before the US is presented. This eyelid conditioning requires the cerebellum (McCormick & Thompson, 1984). The CS and US are conveyed to the cerebellar cortex through the pons → mossy fibers → granule cells → parallel fiber pathway and the inferior olive → climbing fiber pathway (Hesslow, Svensson, & Ivarsson, 1999; Mauk, Steinmetz, & Thompson, 1986; Steinmetz, Lavond, & Thompson, 1989). Thus, a repetitive presentation of CS–US pairing is considered to be equivalent to a repetitive concurrent activation of parallel fibers and climbing fibers, which can induce LTD at parallel fiber–Purkinje cell synapses (Ito & Kano, 1982; Ito et al., 1982). Although it remains elusive whether LTD is indispensable for eyelid conditioning (Schonewille et al., 2011; Welsh et al., 2005), an intact cerebellar cortex is necessary for both acquisition (Garcia, Steele, & Mauk, 1999) and extinction (Perrett & Mauk, 1995) of conditioned responses, and many previous studies suggest the involvement of LTD in this simple form of motor learning.

Anderson et al. (1999) examined how eyelid conditioning affects the morphology of Purkinje cells and found that the complexity of Purkinje cell dendrites was significantly reduced in lobule HVI of conditioned animals (CS–US paired). In those animals, the lengths of total dendritic tree, primary branches, and higher-order branches were shorter, and the total number of higher-order branches was smaller than those in naive animals (Anderson et al., 1999). However, interpretation of this result is not straightforward because a similar degree of dendritic atrophy was also observed in animals that received the same amount of the CS and US without pairing them (CS–US unpaired). The CS–US unpaired animals cannot learn that the CS is followed by the US; hence, the conditioned response is not established (Anderson et al., 1999). Therefore, although the repetitive activation of parallel fibers and climbing fibers could have a long-term effect on the morphology of Purkinje cells, this effect is not uniquely related to either the conjunctive activation of the CS and US or behavioral learning.

Connor et al. (2009) recently revisited the same question: does eyelid conditioning induce structural changes in Purkinje cells? They used electron microscopy to quantify the number of synapses in the outer half of the molecular layer in control (no CS or US was presented), CS–US unpaired, and CS–US paired animals. They found that the number of inhibitory synapses was similar across the groups, but the number of excitatory synapses was significantly fewer in the CS–US paired group than in the control and CS–US unpaired groups (Connor et al., 2009). Since excitatory synapses in the outer half of the molecular layer are mostly formed between parallel fibers and Purkinje cells (Anderson, Alcantara, & Greenough, 1996; Ito, 1984), these results indicate that CS–US pairings result in the reduction of parallel fiber–Purkinje cell synapses.

It is unclear whether the dendritic atrophy observed by Anderson et al. (1999) was also observed in this study, because Connor et al. (2009) only mentioned that the molecular layer thickness and Purkinje cell density did not significantly differ between the groups. However, if the molecular layer thickness was unchanged, it is reasonably speculated that significant dendritic atrophy did not occur. This potential inconsistency might be due to the difference in control animals that were not given either the CS or the US; Anderson et al. (1999) used naive animals, while Connor et al. (2009) used animals that were adapted to the conditioning apparatus.

Unlike the CS–US pairing-independent dendritic atrophy, the CS–US pairing-specific reduction of parallel fiber–Purkinje cell synapses appears to have a direct relevance to the LTD-based model of motor learning in the cerebellum. This point is further discussed in the next section.

2.3. Acrobatic motor skill learning versus eyelid conditioning and functional plasticity versus structural plasticity

Acrobatic learning increases the number of parallel fiber–Purkinje cell synapses, while eyelid conditioning decreases it. Do the underlying mechanisms of learning totally differ between the two learning paradigms? Ohyama, Nores, Murphy, and Mauk (2003) proposed that the cerebellum performs predictive feedforward control of movements (Ohyama et al., 2003). In the case of eyelid conditioning, presentation of unexpected air puff to the eye allows the cerebellum to predict when the next air puff will come and send motor command to elicit precisely timed eyelid closure. Similar types of feedforward control, driven by previous errors, are likely used to improve movement accuracy in acrobatic learning as well. However, unlike eyelid conditioning, various elevated obstacles used in acrobatic learning provide animals with a significantly novel sensory–motor environment. It might be possible to speculate that (i) predictive feedforward control of movements is associated with the reduction of parallel fiber–Purkinje cell synapses and (ii) novel sensory–motor environment increases the number of these synapses. If this is the case, the increase of parallel fiber–Purkinje cell synapses in acrobatic learning animals could be considered as the sum of the increase (by novel sensory–motor environment) and the decrease (by feedforward control of movements) of these synapses, and the increase dominates the decrease. On the other hand, eyelid conditioning is associated with the reduction of these synapses because predictive feedforward control of movement is the major component of this behavior.

In the LTD-based model of eyelid conditioning, LTD at parallel fiber–Purkinje cell synapses results in the reduction of Purkinje cell inhibition onto the deep cerebellar nuclei. This disinhibition allows for the expression of a conditioned response (Heiney, Kim, Augustine, & Medina, 2014). As described above, eyelid conditioning is associated with the reduction in the number of parallel fiber–Purkinje cell synapses, which has the same functional consequence as LTD on the deep cerebellar nuclei. One might therefore assume that LTD eventually leads to the shrinkage and elimination of these synapses. This attractive idea, functional plasticity is eventually transformed into long-lasting structural changes, is recently supported by studies that quantified the number of synaptic AMPA-type glutamate receptors (AMPARs) after adaptation of horizontal optokinetic response (HOKR) (Aziz et al., 2014; Wang et al., 2014). Adaptation of HOKR is a simple form of cerebellar motor learning. Wang et al. (2014) used quantitative SDS-digested freeze-fracture replica labeling and showed that the number of postsynaptic AMPARs at parallel fiber–Purkinje cell synapses decreased after 1 h of HOKR training (Wang et al., 2014). Since LTD at this synapse is expressed as the reduction in the number of postsynaptic AMPARs (Linden, 2001; Linden, Dickinson, Smeyne, & Connor, 1991; Wang &

Linden, 2000), their result strongly indicates that LTD is induced by HOKR training. This reduction of AMPARs was transient: it recovered to basal levels within 24 h. Repetitive training of HOKR over several days did not induce further reduction of synaptic AMPARs, but caused gradual and persistent synapse elimination. These findings provide two crucially important insights into cerebellar learning: (i) LTD may indeed occur at parallel fiber–Purkinje cell synapses during learning under normal circumstances, and (ii) short-term functional plasticity (LTD) might be transformed to long-term structural plasticity (synapse elimination) during the formation of long-term memory.

3. STRUCTURAL PLASTICITY OTHER THAN PARALLEL FIBER–PURKINJE CELL SYNAPSES

3.1. Mossy fiber–Golgi cell synapses

So far, this chapter has focused on parallel fiber–Purkinje cell synapses, but there are other synapses in the cerebellum that are known to show learning-induced structural plasticity. One of such synapses is that of the mossy fiber innervation to Golgi cells. Mossy fibers make large terminal swellings called "rosettes" and provide excitatory inputs to the granule cell layer of the cerebellar cortex (Palay & Chan-Palay, 1974). A mossy fiber rosette consists of a large core structure that is often associated with short filopodia. The large core structure forms synapses onto the dendrites of excitatory granule cells, while the filopodia form synapses onto the dendrites of inhibitory Golgi cells. Thus, a single rosette can be involved in both feedforward excitation and feedforward inhibition in the cerebellar cortex. By using long-term *in vivo* time-lapse microscopy, we recently showed that the large core structure of mossy fiber rosettes was extremely stable over many months without any experimental manipulation to animals (Nishiyama, Colonna, Shen, Carrillo, & Nishiyama, 2014). On the other hand, Ruediger et al. (2011) showed that the filopodia exhibited reversible, learning-associated structural plasticity. They found that cued fear conditioning and rotarod induced robust increases in the number and length of mossy fiber filopodia, indicating that feedforward inhibitory connectivity was strengthened by learning (Ruediger et al., 2011). The role of this selective strengthening of feedforward inhibition in cerebellar learning is unknown. However, similar structural plasticity is also observed in mossy fiber terminals in hippocampal CA3, and it was required for the precision of hippocampus-dependent memories (Ruediger et al., 2011). To examine the role of structural plasticity at mossy fiber–Golgi cell synapses in the cerebellum, it would be instructive to analyze learning-induced sprouting and extension of mossy fiber filopodia by long-term *in vivo* time-lapse microscopy.

3.2. Mossy fiber–deep cerebellar nuclei

Experimental and computational studies indicate that cerebellar-dependent motor learning is mediated by plasticity at parallel fiber–Purkinje cell synapses and mossy fiber–deep cerebellar nuclei synapses (Mauk, 1997; Ohyama, Nores, Medina, Riusech, & Mauk, 2006; Raymond et al., 1996). Like parallel fiber–Purkinje cell synapses, mossy fiber–nucleus synapses can show postsynaptically induced LTP and LTD (Pugh & Raman, 2006; Zhang & Linden, 2006). Kleim et al. (2002) quantified the number of synapses in the nucleus interpositus after 5 days of eyelid conditioning and found that the number of excitatory, but not inhibitory, synapses was significantly increased in CS–US paired animals when compared with CS–US unpaired controls (Kleim et al., 2002). Weeks et al. (2007) conducted similar experiments but performed the quantitative analysis at a relatively earlier phase of eyelid conditioning than Kleim et al. (2002). They found no difference in the synapse number between CS–US paired and CS–US unpaired animals, but the length of postsynaptic density at excitatory synapses was longer in CS–US paired animals, suggesting that the size of excitatory synapses was selectively increased (Weeks et al., 2007). These studies collectively suggest that eyelid conditioning is associated with multiple phases of structural plasticity in the nucleus interpositus; preexisting excitatory synapses are enlarged in the early phase and synaptogenesis forms more excitatory synapses in the later phase.

It should be noted that excitatory inputs in the nucleus largely consist of mossy fibers (carrying CS) and inputs from the inferior olive (carrying US). It is crucially important to determine which signal (CS vs. US) is strengthened in the nucleus through the structural plasticity described above. Experimental data and computer simulation indicate that CS pathway is strengthened in the nucleus (Medina & Mauk, 1999; Ohyama et al., 2006; Steinmetz et al., 1989; Tracy, Thompson, Krupa, & Thompson, 1998); thus, the structural plasticity in excitatory synapses is most likely to occur at mossy fiber inputs. This possibility was recently confirmed by Boele, Koekkoek, De Zeeuw, & Ruigrok (2013). Mossy fibers originating from the lateral part of the basilar pontine nuclei (BPN) relay auditory CS to the cerebellum (Halverson & Freeman, 2010; Kandler & Herbert, 1991). In naive animals, this projection is very sparse in the lateral part of the anterior interposed nucleus (AIN) and dorsolateral hump (DLH), subregions in the cerebellar nuclei that are involved in eyelid conditioning. Boele et al. (2013) showed that the mossy fiber inputs from the lateral BPN significantly increased in the AIN and DLH in CS–US paired animals. Importantly, such an increase was not observed in CS–US unpaired animals. Furthermore, even in CS–US paired animals, mossy fibers originating from the medial BPN (nonauditory part of the BPN) were unchanged. These results indicate that eyelid conditioning strengthens CS pathway to the nucleus by *de novo* formation of mossy fiber inputs carrying CS. Since the nucleus contains a mixture of excitatory and inhibitory neurons, future studies must address whether the postsynaptic targets of these learning-induced new mossy fiber inputs are excitatory or inhibitory.

3.3. Purkinje fiber, Mossy fiber, and olivary inputs to the deep cerebellar nuclei

Despite the many lines of evidence for experience-dependent structural plasticity, it remains largely unknown how external stimuli affect cellular and molecular mechanisms that drive structural changes in synapses. Important insights into this point were recently obtained at both inhibitory (from Purkinje cells) and excitatory (from mossy fibers and olivocerebellar axons) inputs to the deep cerebellar nuclei. Foscarin et al. (2011) housed adult mice either in enriched environment (containing toys, tubes, and running wheels) or in standard cages for one month and compared the size and number of presynaptic terminals on nucleus neurons between the enriched environment group and the standard cage control (Foscarin et al., 2011). They found that the size of Purkinje cell axon terminals and the density of excitatory axon terminals (mixture of mossy fibers and olivocerebellar axons) were significantly increased in the enriched environment group.

The perikarya of nucleus neurons are enwrapped by meshwork of extracellular matrix proteins, called perineuronal nets (Carulli et al., 2006). Perineuronal nets are found in some neuronal populations and thought to restrict neuronal plasticity in the adult brain (Corvetti & Rossi, 2005; Gogolla, Caroni, Luthi, & Herry, 2009; Pizzorusso et al., 2002). In the deep cerebellar nuclei, enriched environment significantly reduces the fraction of neurons that are enwrapped by perineuronal nets (from ~100% in control to ~80% in enriched environment group). Although many neurons were still enwrapped, the degree of enwrapping in those neurons was less than that in the control group (Foscarin et al., 2011). These changes occurred because enriched environment enhanced the activity of matrix-degrading enzymes and reduced the expression of mRNAs that code for key matrix components in perineuronal nets (Foscarin et al., 2011). Since the matrix components in the nets have growth-inhibitory properties (Carulli et al., 2006), it is highly likely that the reduction of perineuronal nets removes local inhibitory cues, which causes the growth of presynaptic inputs in the deep cerebellar nuclei.

Perineuronal nets appear to regulate activity-dependent structural plasticity in other brain regions as well (Miyata, Akagi, Hayashi, Watanabe, & Oohira, 2004; Morita, Oohira, & Miyata, 2010; Sale et al., 2007). It is important to examine whether activity-dependent changes in extracellular matrix proteins regulate structural plasticity regardless of whether synapses are associated with perineuronal nets or not.

4. CONCLUSION

Compared with functional plasticity, the roles of structural plasticity in cerebellar learning have been much less understood. But a relatively small number of studies have revealed learning-induced structural changes in synapses in several different sites in cerebellar circuitry, including two most crucial sites for learning, parallel fiber–Purkinje cell synapses and mossy fiber–deep cerebellar nuclei synapses. Among those studies, the global increase of parallel fiber–Purkinje cell synapses in acrobatic learning is difficult to interpret because of the complexity of the task. Nevertheless, recent studies revealed that, at the level of local circuitry, structural plasticity of both parallel fiber boutons and Purkinje cell spines allowed selective strengthening and weakening of individual synapses, which probably underlies the improvement in complex motor skills. In eyelid conditioning, learning is associated with the decrease in parallel fiber–Purkinje cell synapses and the increase in mossy fiber–nucleus synapses. The direction of these structural changes (decrease vs. increase of synapses) is consistent with the direction of learning-associated functional plasticity (LTD vs. LTP) at both of these synapses. This consistency lets us think that (i) functional plasticity and structural plasticity are regulated by the same mechanisms and (ii) functional plasticity at an early phase of learning is later transformed to structural plasticity to form stable long-term memory. Although the latter idea is supported at least in the case of HOKR adaptation, it remains largely unclear how functional plasticity and structural plasticity relate to each other. Functional plasticity in brain slices is generally induced over a timescale of minutes to hours, whereas behaviorally induced structural plasticity is induced over days to weeks. Because of this significant gap in the timescale, the link between functional plasticity and long-lasting structural plasticity is difficult to investigate. Perhaps, the rule of plasticity induction might be very similar between functional plasticity and structural plasticity (e.g., like LTD induction, conjunctive activation of parallel fibers and climbing fibers may lead to the reduction of parallel fiber–Purkinje cell synapses). But the similar induction mechanism does not necessarily mean that the expression mechanism is the same. It might be possible to speculate that functional plasticity and structural plasticity have different expression mechanisms; thus, structural plasticity can be induced even when functional plasticity is selectively blocked by pharmacological and genetics-based approaches. If this is the case, learning may still occur without measurable LTD. The relationship between functional plasticity and structural plasticity might be a key to explain the ongoing controversy regarding the roles of LTD in cerebellar learning.

ACKNOWLEDGMENTS

I thank Elise Shen, Jennifer Carrillo, and Matasha Dhar for their helpful comments on this manuscript.

REFERENCES

Anderson, B. J., Alcantara, A. A., & Greenough, W. T. (1996). Motor-skill learning: Changes in synaptic organization of the rat cerebellar cortex. *Neurobiology of Learning and Memory*, *66*(2), 221–229.

Anderson, B. J., Relucio, K., Haglund, K., Logan, C., Knowlton, B., Thompson, J., et al. (1999). Effects of paired and unpaired eye-blink conditioning on Purkinje cell morphology. *Learning & Memory*, *6*(2), 128–137.

Aziz, W., Wang, W., Kesaf, S., Mohamed, A. A., Fukazawa, Y., & Shigemoto, R. (2014). Distinct kinetics of synaptic structural plasticity, memory formation, and memory decay in massed and spaced learning. *Proceedings of the National Academy of Sciences of the United States of America*, *111*(1), E194–E202.

Bennett, E. L., Diamond, M. C., Krech, D., & Rosenzweig, M. R. (1964). Chemical and anatomical plasticity brain. *Science*, *146*(3644), 610–619.

Black, J. E., Isaacs, K. R., Anderson, B. J., Alcantara, A. A., & Greenough, W. T. (1990). Learning causes synaptogenesis, whereas motor activity causes angiogenesis, in cerebellar cortex of adult rats. *Proceedings of the National Academy of Sciences of the United States of America*, *87*(14), 5568–5572.

Bliss, T. V., & Lomo, T. (1973). Long-lasting potentiation of synaptic transmission in the dentate area of the anaesthetized rabbit following stimulation of the perforant path. *Journal of Physiology*, *232*(2), 331–356.

Boele, H. J., Koekkoek, S. K., De Zeeuw, C. I., & Ruigrok, T. J. (2013). Axonal sprouting and formation of terminals in the adult cerebellum during associative motor learning. *Journal of Neuroscience*, *33*(45), 17897–17907.

Carey, M. R. (2011). Synaptic mechanisms of sensorimotor learning in the cerebellum. *Current Opinion in Neurobiology*, *21*(4), 609–615.

Carrillo, J., Cheng, S. Y., Ko, K. W., Jones, T. A., & Nishiyama, H. (2013). The long-term structural plasticity of cerebellar parallel fiber axons and its modulation by motor learning. *Journal of Neuroscience*, *33*(19), 8301–8307.

Carulli, D., Rhodes, K. E., Brown, D. J., Bonnert, T. P., Pollack, S. J., Oliver, K., et al. (2006). Composition of perineuronal nets in the adult rat cerebellum and the cellular origin of their components. *Journal of Comparative Neurology*, *494*(4), 559–577.

Chang, F. L., & Greenough, W. T. (1982). Lateralized effects of monocular training on dendritic branching in adult split-brain rats. *Brain Research*, *232*(2), 283–292.

Coesmans, M., Weber, J. T., De Zeeuw, C. I., & Hansel, C. (2004). Bidirectional parallel fiber plasticity in the cerebellum under climbing fiber control. *Neuron*, *44*(4), 691–700.

Connor, S., Bloomfield, J., LeBoutillier, J. C., Thompson, R. F., Petit, T. L., & Weeks, A. C. (2009). Eyeblink conditioning leads to fewer synapses in the rabbit cerebellar cortex. *Behavioral Neuroscience*, *123*(4), 856–862.

Corvetti, L., & Rossi, F. (2005). Degradation of chondroitin sulfate proteoglycans induces sprouting of intact Purkinje axons in the cerebellum of the adult rat. *Journal of Neuroscience*, *25*(31), 7150–7158.

De Paola, V., Holtmaat, A., Knott, G., Song, S., Wilbrecht, L., Caroni, P., et al. (2006). Cell type-specific structural plasticity of axonal branches and boutons in the adult neocortex. *Neuron*, *49*(6), 861–875.

Eccles, J. C., Ito, M., & Szentagothai, J. (1967). *The cerebellum as a neuronal machine*. New York, NY: Springer.

Engert, F., & Bonhoeffer, T. (1999). Dendritic spine changes associated with hippocampal long-term synaptic plasticity. *Nature*, *399*(6731), 66–70.

Federmeier, K. D., Kleim, J. A., & Greenough, W. T. (2002). Learning-induced multiple synapse formation in rat cerebellar cortex. *Neuroscience Letters*, *332*(3), 180–184.

Foscarin, S., Ponchione, D., Pajaj, E., Leto, K., Gawlak, M., Wilczynski, G. M., et al. (2011). Experience-dependent plasticity and modulation of growth regulatory molecules at central synapses. *PLoS One*, *6*(1), e16666.

Gao, Z., van Beugen, B. J., & De Zeeuw, C. I. (2012). Distributed synergistic plasticity and cerebellar learning. *Nature Reviews Neuroscience*, *13*(9), 619–635.

Garcia, K. S., Steele, P. M., & Mauk, M. D. (1999). Cerebellar cortex lesions prevent acquisition of conditioned eyelid responses. *Journal of Neuroscience*, *19*(24), 10940–10947.

Geinisman, Y., Berry, R. W., Disterhoft, J. F., Power, J. M., & Van der Zee, E. A. (2001). Associative learning elicits the formation of multiple-synapse boutons. *Journal of Neuroscience*, *21*(15), 5568–5573.

Gogolla, N., Caroni, P., Luthi, A., & Herry, C. (2009). Perineuronal nets protect fear memories from erasure. *Science*, *325*(5945), 1258–1261.

Greenough, W. T., Larson, J. R., & Withers, G. S. (1985). Effects of unilateral and bilateral training in a reaching task on dendritic branching of neurons in the rat motor-sensory forelimb cortex. *Behavioral and Neural Biology*, *44*(2), 301–314.

Grutzendler, J., Kasthuri, N., & Gan, W. B. (2002). Long-term dendritic spine stability in the adult cortex. *Nature*, *420*(6917), 812–816.

Halverson, H. E., & Freeman, J. H. (2010). Medial auditory thalamic input to the lateral pontine nuclei is necessary for auditory eyeblink conditioning. *Neurobiology of Learning and Memory*, *93*(1), 92–98.

Hansel, C., Linden, D. J., & D'Angelo, E. (2001). Beyond parallel fiber LTD: The diversity of synaptic and non-synaptic plasticity in the cerebellum. *Nature Neuroscience*, *4*(5), 467–475.

Heiney, S. A., Kim, J., Augustine, G. J., & Medina, J. F. (2014). Precise control of movement kinematics by optogenetic inhibition of Purkinje cell activity. *Journal of Neuroscience*, *34*(6), 2321–2330.

Hesslow, G., Svensson, P., & Ivarsson, M. (1999). Learned movements elicited by direct stimulation of cerebellar mossy fiber afferents. *Neuron*, *24*(1), 179–185.

Holtmaat, A. J., Trachtenberg, J. T., Wilbrecht, L., Shepherd, G. M., Zhang, X., Knott, G. W., et al. (2005). Transient and persistent dendritic spines in the neocortex in vivo. *Neuron, 45*(2), 279–291.

Holtmaat, A., Wilbrecht, L., Knott, G. W., Welker, E., & Svoboda, K. (2006). Experience-dependent and cell-type-specific spine growth in the neocortex. *Nature, 441*(7096), 979–983.

Ito, M. (1984). *The cerebellum and neural control.* New York, NY: Raven Press.

Ito, M. (2001). Cerebellar long-term depression: Characterization, signal transduction, and functional roles. *Physiological Reviews, 81*(3), 1143–1195.

Ito, M., & Kano, M. (1982). Long-lasting depression of parallel fiber-Purkinje cell transmission induced by conjunctive stimulation of parallel fibers and climbing fibers in the cerebellar cortex. *Neuroscience Letters, 33*(3), 253–258.

Ito, M., Sakurai, M., & Tongroach, P. (1982). Climbing fibre induced depression of both mossy fibre responsiveness and glutamate sensitivity of cerebellar Purkinje cells. *Journal of Physiology, 324*, 113–134.

Jones, T. A., Chu, C. J., Grande, L. A., & Gregory, A. D. (1999). Motor skills training enhances lesion-induced structural plasticity in the motor cortex of adult rats. *Journal of Neuroscience, 19*(22), 10153–10163.

Jones, T. A., Klintsova, A. Y., Kilman, V. L., Sirevaag, A. M., & Greenough, W. T. (1997). Induction of multiple synapses by experience in the visual cortex of adult rats. *Neurobiology of Learning and Memory, 68*(1), 13–20.

Jorntell, H., & Hansel, C. (2006). Synaptic memories upside down: Bidirectional plasticity at cerebellar parallel fiber-Purkinje cell synapses. *Neuron, 52*(2), 227–238.

Kandler, K., & Herbert, H. (1991). Auditory projections from the cochlear nucleus to pontine and mesencephalic reticular nuclei in the rat. *Brain Research, 562*(2), 230–242.

Kleim, J. A., Freeman, J. H., Jr., Bruneau, R., Nolan, B. C., Cooper, N. R., Zook, A., et al. (2002). Synapse formation is associated with memory storage in the cerebellum. *Proceedings of the National Academy of Sciences of the United States of America, 99*(20), 13228–13231.

Kleim, J. A., Swain, R. A., Armstrong, K. A., Napper, R. M., Jones, T. A., & Greenough, W. T. (1998). Selective synaptic plasticity within the cerebellar cortex following complex motor skill learning. *Neurobiology of Learning and Memory, 69*(3), 274–289.

Knott, G. W., Quairiaux, C., Genoud, C., & Welker, E. (2002). Formation of dendritic spines with GABAergic synapses induced by whisker stimulation in adult mice. *Neuron, 34*(2), 265–273.

Lee, W. C., Huang, H., Feng, G., Sanes, J. R., Brown, E. N., So, P. T., et al. (2006). Dynamic remodeling of dendritic arbors in GABAergic interneurons of adult visual cortex. *PLoS Biology, 4*(2), e29.

Lee, K. J., Jung, J. G., Arii, T., Imoto, K., & Rhyu, I. J. (2007). Morphological changes in dendritic spines of Purkinje cells associated with motor learning. *Neurobiology of Learning and Memory, 88*(4), 445–450.

Lee, K. J., Park, I. S., Kim, H., Greenough, W. T., Pak, D. T., & Rhyu, I. J. (2013). Motor skill training induces coordinated strengthening and weakening between neighboring synapses. *Journal of Neuroscience, 33*(23), 9794–9799.

Lendvai, B., Stern, E. A., Chen, B., & Svoboda, K. (2000). Experience-dependent plasticity of dendritic spines in the developing rat barrel cortex in vivo. *Nature, 404*(6780), 876–881.

Lev-Ram, V., Wong, S. T., Storm, D. R., & Tsien, R. Y. (2002). A new form of cerebellar long-term potentiation is postsynaptic and depends on nitric oxide but not cAMP. *Proceedings of the National Academy of Sciences of the United States of America, 99*(12), 8389–8393.

Linden, D. J. (2001). The expression of cerebellar LTD in culture is not associated with changes in AMPA-receptor kinetics, agonist affinity, or unitary conductance. *Proceedings of the National Academy of Sciences of the United States of America, 98*(24), 14066–14071.
Linden, D. J., Dickinson, M. H., Smeyne, M., & Connor, J. A. (1991). A long-term depression of AMPA currents in cultured cerebellar Purkinje neurons. *Neuron, 7*(1), 81–89.
Majewska, A. K., Newton, J. R., & Sur, M. (2006). Remodeling of synaptic structure in sensory cortical areas in vivo. *Journal of Neuroscience, 26*(11), 3021–3029.
Matsuzaki, M., Honkura, N., Ellis-Davies, G. C., & Kasai, H. (2004). Structural basis of long-term potentiation in single dendritic spines. *Nature, 429*(6993), 761–766.
Mauk, M. D. (1997). Roles of cerebellar cortex and nuclei in motor learning: Contradictions or clues? *Neuron, 18*(3), 343–346.
Mauk, M. D., Garcia, K. S., Medina, J. F., & Steele, P. M. (1998). Does cerebellar LTD mediate motor learning? Toward a resolution without a smoking gun. *Neuron, 20*(3), 359–362.
Mauk, M. D., Steinmetz, J. E., & Thompson, R. F. (1986). Classical conditioning using stimulation of the inferior olive as the unconditioned stimulus. *Proceedings of the National Academy of Sciences of the United States of America, 83*(14), 5349–5353.
McCormick, D. A., & Thompson, R. F. (1984). Cerebellum: Essential involvement in the classically conditioned eyelid response. *Science, 223*(4633), 296–299.
Medina, J. F., & Mauk, M. D. (1999). Simulations of cerebellar motor learning: Computational analysis of plasticity at the mossy fiber to deep nucleus synapse. *Journal of Neuroscience, 19*(16), 7140–7151.
Miyata, S., Akagi, A., Hayashi, N., Watanabe, K., & Oohira, A. (2004). Activity-dependent regulation of a chondroitin sulfate proteoglycan 6B4 phosphacan/RPTPbeta in the hypothalamic supraoptic nucleus. *Brain Research, 1017*(1–2), 163–171.
Morita, S., Oohira, A., & Miyata, S. (2010). Activity-dependent remodeling of chondroitin sulfate proteoglycans extracellular matrix in the hypothalamo-neurohypophysial system. *Neuroscience, 166*(4), 1068–1082.
Nagerl, U. V., Eberhorn, N., Cambridge, S. B., & Bonhoeffer, T. (2004). Bidirectional activity-dependent morphological plasticity in hippocampal neurons. *Neuron, 44*(5), 759–767.
Nishiyama, N., Colonna, J., Shen, E., Carrillo, J., & Nishiyama, H. (2014). Long-term in vivo time-lapse imaging of synapse development and plasticity in the cerebellum. *Journal of Neurophysiology, 111*(1), 208–216.
Nishiyama, H., Fukaya, M., Watanabe, M., & Linden, D. J. (2007). Axonal motility and its modulation by activity are branch-type specific in the intact adult cerebellum. *Neuron, 56*(3), 472–487.
Ohyama, T., Nores, W. L., Medina, J. F., Riusech, F. A., & Mauk, M. D. (2006). Learning-induced plasticity in deep cerebellar nucleus. *Journal of Neuroscience, 26*(49), 12656–12663.
Ohyama, T., Nores, W. L., Murphy, M., & Mauk, M. D. (2003). What the cerebellum computes. *Trends in Neurosciences, 26*(4), 222–227.
Okamoto, K., Nagai, T., Miyawaki, A., & Hayashi, Y. (2004). Rapid and persistent modulation of actin dynamics regulates postsynaptic reorganization underlying bidirectional plasticity. *Nature Neuroscience, 7*(10), 1104–1112.
Ostroff, L. E., Fiala, J. C., Allwardt, B., & Harris, K. M. (2002). Polyribosomes redistribute from dendritic shafts into spines with enlarged synapses during LTP in developing rat hippocampal slices. *Neuron, 35*(3), 535–545.
Palay, S., & Chan-Palay, V. (1974). *Cerebellar cortex: Cytology and organization.* New York, NY: Springer.
Perrett, S. P., & Mauk, M. D. (1995). Extinction of conditioned eyelid responses requires the anterior lobe of cerebellar cortex. *Journal of Neuroscience, 15*(3 Pt. 1), 2074–2080.

Pichitpornchai, C., Rawson, J. A., & Rees, S. (1994). Morphology of parallel fibres in the cerebellar cortex of the rat: An experimental light and electron microscopic study with biocytin. *Journal of Comparative Neurology*, *342*(2), 206–220.

Pizzorusso, T., Medini, P., Berardi, N., Chierzi, S., Fawcett, J. W., & Maffei, L. (2002). Reactivation of ocular dominance plasticity in the adult visual cortex. *Science*, *298*(5596), 1248–1251.

Popov, V. I., & Bocharova, L. S. (1992). Hibernation-induced structural changes in synaptic contacts between mossy fibres and hippocampal pyramidal neurons. *Neuroscience*, *48*(1), 53–62.

Popov, V. I., Bocharova, L. S., & Bragin, A. G. (1992). Repeated changes of dendritic morphology in the hippocampus of ground squirrels in the course of hibernation. *Neuroscience*, *48*(1), 45–51.

Pugh, J. R., & Raman, I. M. (2006). Potentiation of mossy fiber EPSCs in the cerebellar nuclei by NMDA receptor activation followed by postinhibitory rebound current. *Neuron*, *51*(1), 113–123.

Raymond, J. L., Lisberger, S. G., & Mauk, M. D. (1996). The cerebellum: A neuronal learning machine? *Science*, *272*(5265), 1126–1131.

Ruediger, S., Vittori, C., Bednarek, E., Genoud, C., Strata, P., Sacchetti, B., et al. (2011). Learning-related feedforward inhibitory connectivity growth required for memory precision. *Nature*, *473*(7348), 514–518.

Sale, A., Maya Vetencourt, J. F., Medini, P., Cenni, M. C., Baroncelli, L., De Pasquale, R., et al. (2007). Environmental enrichment in adulthood promotes amblyopia recovery through a reduction of intracortical inhibition. *Nature Neuroscience*, *10*(6), 679–681.

Schonewille, M., Gao, Z., Boele, H. J., Veloz, M. F., Amerika, W. E., Simek, A. A., et al. (2011). Reevaluating the role of LTD in cerebellar motor learning. *Neuron*, *70*(1), 43–50.

Steinmetz, J. E., Lavond, D. G., & Thompson, R. F. (1989). Classical conditioning in rabbits using pontine nucleus stimulation as a conditioned stimulus and inferior olive stimulation as an unconditioned stimulus. *Synapse*, *3*(3), 225–233.

Stettler, D. D., Yamahachi, H., Li, W., Denk, W., & Gilbert, C. D. (2006). Axons and synaptic boutons are highly dynamic in adult visual cortex. *Neuron*, *49*(6), 877–887.

Trachtenberg, J. T., Chen, B. E., Knott, G. W., Feng, G., Sanes, J. R., Welker, E., et al. (2002). Long-term in vivo imaging of experience-dependent synaptic plasticity in adult cortex. *Nature*, *420*(6917), 788–794.

Tracy, J. A., Thompson, J. K., Krupa, D. J., & Thompson, R. F. (1998). Evidence of plasticity in the pontocerebellar conditioned stimulus pathway during classical conditioning of the eyeblink response in the rabbit. *Behavioral Neuroscience*, *112*(2), 267–285.

Wang, Y. T., & Linden, D. J. (2000). Expression of cerebellar long-term depression requires postsynaptic clathrin-mediated endocytosis. *Neuron*, *25*(3), 635–647.

Wang, W., Nakadate, K., Masugi-Tokita, M., Shutoh, F., Aziz, W., Tarusawa, E., et al. (2014). Distinct cerebellar engrams in short-term and long-term motor learning. *Proceedings of the National Academy of Sciences of the United States of America*, *111*(1), E188–E193.

Weeks, A. C., Connor, S., Hinchcliff, R., LeBoutillier, J. C., Thompson, R. F., & Petit, T. L. (2007). Eye-blink conditioning is associated with changes in synaptic ultrastructure in the rabbit interpositus nuclei. *Learning & Memory*, *14*(6), 385–389.

Welsh, J. P., Yamaguchi, H., Zeng, X. H., Kojo, M., Nakada, Y., Takagi, A., et al. (2005). Normal motor learning during pharmacological prevention of Purkinje cell long-term depression. *Proceedings of the National Academy of Sciences of the United States of America*, *102*(47), 17166–17171.

Woolley, C. S., Gould, E., Frankfurt, M., & McEwen, B. S. (1990). Naturally occurring fluctuation in dendritic spine density on adult hippocampal pyramidal neurons. *Journal of Neuroscience*, *10*(12), 4035–4039.

Xu, T., Yu, X., Perlik, A. J., Tobin, W. F., Zweig, J. A., Tennant, K., et al. (2009). Rapid formation and selective stabilization of synapses for enduring motor memories. *Nature*, *462*(7275), 915–919.

Yang, G., Pan, F., & Gan, W. B. (2009). Stably maintained dendritic spines are associated with lifelong memories. *Nature*, *462*(7275), 920–924.

Zhang, W., & Linden, D. J. (2006). Long-term depression at the mossy fiber-deep cerebellar nucleus synapse. *Journal of Neuroscience*, *26*(26), 6935–6944.

Zhou, Q., Homma, K. J., & Poo, M. M. (2004). Shrinkage of dendritic spines associated with long-term depression of hippocampal synapses. *Neuron*, *44*(5), 749–757.

Zuo, Y., Lin, A., Chang, P., & Gan, W. B. (2005). Development of long-term dendritic spine stability in diverse regions of cerebral cortex. *Neuron*, *46*(2), 181–189.

Zuo, Y., Yang, G., Kwon, E., & Gan, W. B. (2005). Long-term sensory deprivation prevents dendritic spine loss in primary somatosensory cortex. *Nature*, *436*(7048), 261–265.

CHAPTER TWO

Cerebellar Mechanisms of Learning and Plasticity Revealed by Delay Eyelid Conditioning

Michael D. Mauk*,†,1, Wenke Li*, Andrei Khilkevich*, Hunter Halverson*

*Center for Learning and Memory, The University of Texas, Austin, Texas, USA
†Department of Neuroscience, The University of Texas, Austin, Texas, USA
[1]Corresponding author: e-mail address: mike@mail.clm.utexas.edu

Contents

Abstract

The analysis of well-defined behaviors that require the cerebellum has helped reveal many key mechanisms operating in the cerebellum to mediate learning and feed-forward prediction. These systems include eyelid conditioning, adaptation of the vestibuloocular reflex, smooth pursuit eye movements, and arm-reaching tasks. This review focuses specifically on the variety of findings that have come from the use of eyelid conditioning to study the cerebellum. Phenomena discussed include sites and rules for plasticity, temporal coding and mechanisms of timing, cerebellar control of climbing fibers and its role in bidirectional learning, extinction of conditioned responses, and the phenomenon of savings.

1. INTRODUCTION

The study of delay eyelid conditioning has produced a wealth of information about the properties and mechanisms of learning and plasticity in the

International Review of Neurobiology, Volume 117
ISSN 0074-7742
http://dx.doi.org/10.1016/B978-0-12-420247-4.00002-6

cerebellum. This represents the third-generation emphasis in delay eyelid conditioning research. The first wave began in the early twentieth century with the use of eyelid conditioning in both humans and animals to study the behavioral properties of associative learning. This spans back to the work of Hilgard and Marquis (1940), includes Gormezano's studies using the rabbit nictitating membrane preparation (Gormezano, 1972; Gormezano, Kehoe, & Marshall, 1983; Gormezano, Prokasy, & Thompson, 1987; Gormezano, Schneiderman, Deaux, & Fuentes, 1962), and also includes the enormous corpus of work by Gormezano's student James Kehoe (e.g., Kehoe & Napier, 1991; Millenson, Kehoe, & Gormezano, 1977). Michael Patterson and Richard F. Thompson initiated the second generation of studies by developing the use of delay conditioning as a model system to localize the site(s) of plasticity for a well-behaved form of associative learning in a mammal (e.g., Cegavske, Patterson, & Thompson, 1979). Work from Thompson's lab and other labs established the necessity of cerebellum for delay eyelid conditioning (McCormick & Thompson, 1984). This era produced the foundational findings that cerebellar lesions affect delay eyelid conditioning, that the conditioned stimulus (CS) is conveyed to the cerebellum via mossy fiber inputs, that the unconditioned stimulus (US) is conveyed by climbing fiber inputs, and that cerebellar output drives the premotor and motor neurons responsible for the expression of the conditioned responses. Eyelid conditioning studies of this era accomplished the goal of localizing the plasticity that is responsible for a specific and well-characterized form of learning (Thompson, 1976, 2013). The emergence of these findings was the genesis of the third generation of eyelid conditioning studies. Here, the emphasis has been the use of eyelid conditioning, with the control that it provides over inputs and the relatively simple readout it provides of cerebellar output, to study the mechanisms of learning, plasticity, timing, and information processing in the cerebellum (e.g., Mauk & Donegan, 1997).

In this chapter, we focus on this third-generation application of eyelid conditioning. We wish to state straightaway that it has not been eyelid conditioning studies alone that have revealed so much in the past 30 years about the cerebellar mechanisms of learning, memory, and computation. We fully acknowledge and greatly admire the numerous contributions from labs making using of vestibuloocular reflex, saccades, smooth pursuit, and arm/wrist movements. Our rationale for the emphasis on eyelid conditioning studies is a desire for focus and clarity and on the existence of other fine reviews that have focused specifically on non-eyelid studies (e.g., Boyden, Katoh, & Raymond, 2004).

2. EYELID CONDITIONING AS A TOOL TO STUDY THE CEREBELLUM

The factors that make eyelid conditioning useful for studying the cerebellum are simple: eyelid conditioning provides an experimentally simply way to control cerebellar inputs and to infer an approximation of its output. Behaviorally, eyelid conditioning training involves the paired presentation of a CS, which is usually a tone, and an US that threatens the eye—usually an air puff directed at the cornea or subdermal electrical stimulation near the eye. Before training, the US will elicit a reflex response of the eyelids (they close), whereas the CS has little to no effect on eyelid position. With paired CS + US presentations, the subject will eventually acquire conditioned eyelid responses; the eyelids now close in response to the CS. This learning is bidirectional in that presentation of CS-alone trials to a trained animal will result in the extinction of conditioned responses. Upon retraining with CS + US trials, the animals reacquire conditioned responses in far fewer trials than required for initial acquisition—the phenomenon of savings.

For convenience, we will refer to eyelid responses in this chapter. Many studies have focused on the nictitating membrane, a third inner eyelid that retracts passively across the eye as the eyeball retracts. For these studies, the external eyelids are restrained and the position of the nictitating membrane is measured, whereas for eyelid conditioning, the dependent measure is the position of the external eyelids or essentially the distance between the eyelids. Although an early study demonstrated these measures covary considerably (McCormick, Lavond, & Thompson, 1982), the nictitating membrane response occurs passively as the eyeball contracts and is thus delayed and temporally distorted with respect to eyelid responses. For this reason, the eyelid responses are a closer reflection of motor neuron firing, and therefore, most mechanistically oriented studies, particular those where the timing of the responses is relevant, have used the eyelid measure.[1]

For as simple as eyelid conditioning is to implement, its behavioral properties are surprisingly rich and intricate. Examples include the observations that the rates of acquisition and extinction depend on the interstimulus

[1] We prefer the term "eyelid conditioning" to the more common "eyeblink conditioning" in part because the conditioned responses are not blinks—a word that has specific implications about the response that do not apply to conditioned responses—and in part because "eyeblink" is not an English word.

interval (ISI, the time between CS and US onsets), that the timing of the responses depends on the ISI, and that eyelid conditioning shows robust and varied forms of savings. As we shall discuss later, in many different ways, this has served as a valuable source of insight about underlying cerebellar processes.

More so than any other factor, it is the unusually direct way in which eyelid conditioning engages the cerebellum that enables mechanistic studies. Several reviews have presented the evidence in detail, so here, we will be brief (Mauk & Donegan, 1997; McCormick & Thompson, 1984). The presentation of the CS is conveyed to the cerebellum via activation of mossy fibers that arise in the pontine nuclei. Recordings show the CS activation of these neurons, lesions produce effects equivalent of omitting the CS, and perhaps most importantly, electrical stimulation of mossy fibers can substitute for a tone CS to produce normal learning. Because it does not appear important precisely which mossy fibers are activated, it is relatively straightforward to implement conditioning with a stimulation CS. Indeed, mossy fiber stimulation is often used when precise control over activity of the mossy fiber input is important. Via parallel observations, we know that the US is conveyed by the activation of climbing fiber inputs whose cell bodies are in the dorsal accessory nucleus of the inferior olive. Here, the situation with using stimulation as the US is more complex. There is a somatosensory map representation in the dorsal accessory olive and the particular response supported by the US will depend on the precise placement of the stimulating electrode within this map. On the output side, again stimulation, lesions and recording studies reveal that a particular subset of neurons in the lateral portion of the anterior (lateral?) interpositus (deep) nucleus drive the expression of the conditioning responses.

Here is the opportunity that has been exploited by so many mechanistic studies. Eyelid conditioning provides a simple way to control the activation of mossy fiber and climbing fiber inputs to the cerebellum and, by simply measuring eyelid responses, to have an approximation of the output of the cerebellum via the deep nucleus. This relatively direct relationship between eyelid conditioning and the cerebellum also greatly facilitates computer simulation studies of cerebellum. A wealth of anatomical and physiological data constrain the construction of computer simulations of the cerebellum (e.g., Buonomano & Mauk, 1994; Medina & Mauk, 1999, 2000). Equally importantly, eyelid conditioning data allow the performance of these simulations to be evaluated in biologically relevant and quantitative ways. When these simulations are provided with eyelid conditioning-like inputs (mossy and climbing fiber), the large corpus of behavioral data on eyelid conditioning inform what sort of outputs a successful simulation should produce. These advantages are not available for any other system in the brain, and much progress on cerebellar properties and mechanisms has been made possible by these simple but conceptually important factors.

3. SITES AND RULES FOR LEARNING-RELATED PLASTICITY

Eyelid conditioning studies played a prominent role in identifying that two major sites of plasticity can interact to support cerebellar learning. In the early 1990s, intense debates existed in both the VOR adaptation and eyelid conditioning fields regarding the importance of learning-related plasticity in the cerebellar cortex versus learning-related plasticity in the deep nuclei. Masao Ito and Chris Yeo were champions of plasticity in cerebellar cortex for VOR adaptation and eyelid conditioning, respectively. Steve Lisberger and Dick Thompson were the main proponents for plasticity in the deep nuclei, again, respectively, for VOR adaptation and eyelid conditioning. Although most deep nucleus plasticity proponents assumed the climbing fiber collaterals controlled the induction of plasticity in the deep nucleus, Miles and Lisberger (1981) proposed that the cerebellar cortex computed an error signal and that the Purkinje cell inputs are the teaching signal for the induction of deep nucleus plasticity. This new and specific hypothesis could explain the combination of plasticity in the deep nucleus and the necessity of the cerebellar cortex in cerebellar learning.

It was eyelid conditioning studies that first gave rise to the proposal, first from Perrett, Ruiz, and Mauk (1993) and then from Mauk and Donegan (1997), that plasticity in both the cerebellar cortex and the deep nuclei mediates cerebellar learning. Mauk and Donegan combined a variant of the Miles and Lisberger hypothesis with climbing fiber-controlled plasticity in the cerebellar cortex to explain (1) the necessity of cerebellar cortex for acquisition and extinction of conditioned eyelid responses and (2) the preservation of poorly timed conditioned responses following lesions of the cerebellar cortex in already-trained animals. In the Mauk and Donegan hypothesis, plasticity in the cerebellar cortex is necessary to train Purkinje cells (at least in eyelid conditioning) to elicit learned pauses in activity that can then serve as the teaching input, much like the Miles and Lisberger theory, for the induction of plasticity in the deep nucleus. This explains the inability of animals with cerebellar cortex lesions to acquire conditioned responses (Garcia, Steele, & Mauk, 1999): the lesions remove one site of plasticity and the teaching input for induction of plasticity at the second site. That way, not counting learned timing of the response, whatever responses did or did not exist prior to the lesion would remain following the lesion. This idea was supported by the observation of Perrett and Mauk (1995) that lesions of cerebellar cortex prevent the extinction of the short-latency responses that are unmasked by lesions of cerebellar cortex in already-trained animals. Thus, pretraining lesions of cerebellar cortex prevent learning, whereas posttraining lesions do not abolish responses—because they are mediated by already induced plasticity in the deep nucleus—but these responses present after the lesion cannot be extinguished.

Inspired by the theory of David Marr (1969), Ito and his colleagues discovered the phenomenon of climbing fiber-controlled long-term depression (LTD) at granule cell to Purkinje cell synapses (Ito, 1988; Ito & Kano, 1982; Ito, Sakurai, & Tongroach, 1982). A great deal of subsequent work, much of it from the laboratory of David Linden and colleagues, has identified many of the key molecular mechanisms that mediate the induction and expression of this plasticity. Along the way, it was established that cerebellar LTD is expressed postsynaptically via changes in the expression of AMPA-type glutamate receptors. In 1987, Sakurai demonstrated that the granule-to-Purkinje synapses can also undergo long-term potentiation (LTP) when activated in the absence of the climbing fiber input (Sakurai, 1987). Somewhat confusingly, this form of LTP was shown to be expressed presynaptically. Much later, Lev-Ram, Mehta, Kleinfeld, and Tsien (2003) identified another form of LTP at these synapses that occurs when the synapses are activated in the absence of climbing fiber input that is expressed postsynaptically and that appears to reverse the changes induced by LTD. It is this form of LTP that is believed to be the complement to climbing fiber-induced LTD, while a role for the presynaptically expressed form of LTP remains less clear. Ohyama, Voicu, Kalmbach, and Mauk (2010) proposed that this form of LTP may fade over the course of hours and mediate short-term behavioral changes apparent in eyelid conditioning.

4. THE ROLE OF TIMING IN CEREBELLAR LEARNING

The role of the ISI in eyelid conditioning (the time between the onset of the CS and the onset of the US) reveals two fundamentally important properties of cerebellar learning. Cerebellar learning is associative, as revealed by the limited range of ISIs that support learning (Schneiderman & Gormezano, 1964). A CS (mossy fiber input) must precede an US (climbing fiber input) by 120–150 ms to promote learning. Robust learning occurs with ISIs between 250 and 750 ms. As the ISI increases beyond 750 ms or so, the robustness of conditioning declines until learning does not occur with ISIs beyond 2 s. A recent study also investigated another aspect of the associative properties of cerebellar learning—that is, how soon after the offset of a mossy fiber input does a climbing fiber input have to arrive to support learning? Kalmbach, Ohyama, Kreider, Riusech, and Mauk (2009) used a 500 ms train of mossy fiber stimulation paired with an US presented at different times after train offset. The essential observation was that the robustness of learning declined as this "trace interval" increased until only half the animals showed learning with a 400 ms trace interval and none learned at a 500 ms trace interval. The mechanism by which cerebellar learning can be maintained with a 300–400 ms gap between mossy fiber offset and climbing fiber onset is not known. These data also provide a specific and concrete explanation for why trace eyelid conditioning with trace intervals of 500 and above requires, in addition to the cerebellum, structures such as prefrontal cortex and hippocampus. These forebrain structures have been shown to provide an input to certain mossy fibers that persists through the trace interval to support cerebellar learning. In this sense, the forebrain input turns trace conditioning, from the perspective of the cerebellum, back into delay conditioning where the mossy fiber and climbing fiber inputs overlap in time, which makes cerebellar learning possible.

Within the effective range, the ISI also determines the timing of the conditioned responses. The onset and velocity profile of conditioned responses vary with the ISI so that conditioned responses are timed to peak just before the onset of the US. So, shorter ISIs promote conditioned responses with relatively short latencies to onset and higher eyelid velocity, whereas longer ISIs promote conditioned responses with delayed latencies to onset and lower peak velocities (Mauk & Ruiz, 1992). There also is the general tendency for the variability of the timing of the conditioned responses to increase with the ISI in a Weber-like fashion. Two observations show that this learned timing is a property of the cerebellum and is neither conveyed to the cerebellum via its inputs nor generated by downstream processing. First, it is apparent even when stimulation of mossy fibers is substituted for the CS (e.g., Kalmbach, Davis, Ohyama, Nores, & Mauk, 2010). Because this stimulation is tonic and conveys a minimum of temporal information, the learned timing must be generated in the cerebellum or downstream from it. Second, recordings from deep nucleus neurons show that the bursts of cerebellar output that drive the expression of the conditioned responses are themselves well timed (Choi & Moore, 2003; McCormick & Thompson, 1984). More specifically, the learned timing of conditioned responses appears to be a property of the cerebellar cortex as lesion of cerebellar cortex abolishes timing, defaulting the conditioned responses to a relatively fixed and short latency (Garcia & Mauk, 1998; Kalmbach, Ohyama, & Mauk, 2010; Perrett et al., 1993). Also, Purkinje cell responses during the expression of conditioned responses clearly support the idea that their activity largely determines the timing of conditioned eyelid responses (Halverson, Khilkevich, & Mauk, unpublished).

Computer simulations of the cerebellum as tested by the inputs and behavioral properties of eyelid conditioning have contributed three important computational hypotheses regarding the mechanisms of timing and temporal coding in the cerebellum. First, Buonomano and Mauk (1994) proposed that sparse recurrent interactions between granule cells and Golgi cells give rise to temporal coding. Specifically, the hypothesis was that during a tonic mossy fiber input (as might occur during a tone CS or especially during mossy fiber stimulation used as a CS), different subsets of granule cells become active at different times during the stimulus (see also Mauk & Donegan, 1997). From this, an explanation for conditioned response timing arises from the climbing fiber-controlled induction of LTD that would modify only those granule cells to Purkinje synapses that were active just before US onset. Mauk and Donegan also noted that since learning would require consistency in this temporal pattern from one trial to the next, breakdown in this consistency could explain such phenomena as long CSs losing their ability to support conditioning. Second, computer simulations suggested that differential conditioning within each trial could help sharpen the timing of conditioned responses, particularly at longer ISIs. The basis for this will be covered in Section 5, but the basic idea is that while acquisition processes are increasing the amplitude of the conditioned responses just before US onset, extinction processes are decreasing response amplitude earlier in the trial. This has the effect of increasing the latency to onset and increasing rise time of the conditioned responses. Experimental predictions from this hypothesis were supported by subsequent experiments (Medina, Garcia, Nores, Taylor, & Mauk, 2000). A third more recent computer simulation-generated prediction is that a unique subset of granule cells can become active when a subset of the currently active mossy fiber inputs go quiet. This "temporal subtraction" hypothesis was strongly supported by empirical test (Kalmbach, Voicu, Ohyama, & Mauk, 2011). The computational and functional implications of the learned timing displayed by the cerebellum have been reviewed elsewhere (Ohyama, Medina, Nores, & Mauk, 2002). In recent studies, the timing properties of cerebellar learning have been beautifully demonstrated and experimentally exploited in studies of smooth pursuit (Medina, Carey, & Lisberger, 2005; Medina & Lisberger, 2008).

5. FEEDBACK CONTROL OF CLIMBING FIBER EQUILIBRIUM ACTIVITY

Germund Hesslow and colleagues first showed that the cerebellum inhibits its own climbing fiber input via projections from the deep cerebellar nuclei to the inferior olive (Andersson, Garwicz, & Hesslow, 1988). The fundamental importance of this control in cerebellar learning was not appreciated until attempts were made to build computer simulations of the cerebellum that were tested for their ability to replicate the behavioral features of eyelid conditioning. In the absence of this connection, simulations invariably demonstrate steady drift in the strengths of the granule cell to Purkinje cell synapses until these synapses are all as strong or as weak as they can be. This persistent drift completely precludes normal cerebellar function, let alone any hope for learning (Medina & Mauk, 1999).

The inferior olive neurons that give rise to climbing fibers display fairly regular spontaneous activity around 1 Hz. In the absence of such activity, the simulations showed that granule-to-Purkinje synapses would undergo a steady increase in strength until eventually saturated at full strength. With too much spontaneous activity in climbing fibers, the result was the same only in the direction of decreasing synapse strength. Any such tendency completely precludes any capacity for learning in the cerebellum. For one thing, this produces cerebellar output that is relatively fixed and extreme, either very high or very low. For another, systematic drift of synaptic weights erases all existing cerebellar cortex memories, even if the saturation is eventually reversed.

The simulations demonstrated that what is required is just the right amount of spontaneous activity in climbing fiber inputs and that the tolerances for "just the right amount" are very high (Medina, Nores, & Mauk, 2002). The only way that such precision can be ensured is to have cerebellar output influence climbing fiber activity through inhibition, the very projection Hesslow had demonstrated. This inhibition serves as negative feedback—it increases climbing fiber activity (and therefore biasing plasticity toward LTD) when granule-to-Purkinje synapses drift upward and decreases climbing fiber activity when the overall strength of granule-to-Purkinje synapses drifts down (biasing plasticity toward LTP). From this work, we came to the conclusion that it is simply not possible to build a properly functioning simulation without this property. Nor, by inference, would it be possible for the cerebellum to function properly without cerebellar inhibition of climbing fibers.

Both simulation and experimental results show that cerebellar inhibition of climbing fibers implements another important computational function: it provides the ability to modulate climbing fiber activity below its spontaneous level and therefore signal learning in the opposite direction. In the case of eyelid conditioning, this is the signal that initiates the extinction of conditioned responses. For VOR adaptation, it would be the signal for decreases in gain, and for saccades, it will signal lower amplitude. This signal occurs because the expression of a conditioned response involves a transient increase in activity in the cerebellar deep nucleus neurons. Thus, when there is an increase in cerebellar output (and thus an increase in inhibition of climbing fibers) without an US, then there is a net induction of LTP at the granule-to-Purkinje synapses, which acts to decrease the amplitude and likelihood of a conditioned response. For eyelid conditioning, this process was demonstrated directly by preventing the extinction of conditioned responses via reversible pharmacological block of the inhibitory synapses onto the olivary neurons (Medina et al., 2002). With spontaneous climbing fiber activity regulated by cerebellar output, (1) background activity is driven to an intermediate level that precludes drift of synaptic weights, (2) transient increases in activity above this level are the signal to increase the amplitude of the response (or more generally to increase cerebellar output for that particular mossy fiber input), and (3) transient decreases in climbing fiber activity below this level are the signal to decrease the amplitude of the response.

There are interesting parallels between the behavior of climbing fibers and the dopamine neurons, which are believed to control the induction of plasticity in the striatum. Both have a low level of spontaneous activity, for both transient increases signal learning in one direction and transient decreases signal learning in the other direction. Based on this, we speculate that, like climbing fibers, dopamine neurons are influenced by a projection from some portion of the basal ganglia in a direction that implements negative feedback. This would preclude the wholesale drift of synaptic weights in the striatum and would also mediate the well-characterized decrease in dopamine neuron activity when a reward is expected, but omitted.

6. THE ROLE OF FEEDBACK INHIBITION OF CLIMBING FIBERS IN RESPONSE TIMING

There are also important implications of feedback control of inhibition on the learning of response timing. The combination of climbing fiber spontaneous activity and feedback inhibition of this activity implements this simple rule: when the cerebellum makes a response—through a burst of activity in deep nucleus neurons—when there is no excitatory input to the climbing fibers, the response gets smaller and less likely (it extinguishes). This process appears to operate within trials too in the sense that if there is a time during a CS where the cerebellum is expressing a response in the absence of the US, the subsequent responses at that time during the CS will get smaller. Computer simulations of the cerebellum predict this: by artificially turning of LTP during the CS during training, the conditioned responses that are learned are less well timed. Specifically, the responses begin too early. This means that during a training trial, particularly one with a longer ISI where the conditioned response timing will be delayed, the cerebellum is learning to respond later in the CS and is learning not to respond earlier in the CS. The latter process in particular contributes to sharper conditioned response timing with the latency to onset delayed to make the responses peak near US onset. The results of Medina et al. (2000) provide support for this simulation-based prediction. Small lesions of cerebellar cortex were observed to decrease the latency to onset of the conditioned responses. Over subsequent days of training, this timing deficit often gradually resolved. The authors suggested that the lesions damage some, but not all, of the Purkinje cells responsible for the expression of the conditioned responses. If Purkinje cells are learning to suppress the expression of responses early in the CS, removing some Purkinje cells could unmask shorter latency conditioned responses, as was observed. If the lesion spared enough Purkinje cells to support further learning to suppress responding, then the early responses unmasked by the lesions could be suppressed through further learning. These authors observed the same pattern of results when a subset of the simulation's Purkinje cells were disabled after initial learning.

In summary, we believe it is not possible to have a properly functioning cerebellum without the inhibitory projection from the deep cerebellar nuclei to the source of the climbing fibers in the inferior olive. This inhibitory projection appears to be necessary even for the cerebellum to be in a normal operational mode where it is capable of learning and maintaining the memory of previous learning. The constant drift of synaptic weights that inevitably occurs in the absence of this projection would produce such deficits. The inhibition also provides the basis for the signal to climbing fibers responsible for decreasing responses, such as occurs during extinction training in eyelid conditioning. Finally, even the proper adaptive timing of the conditioned responses is influenced by the ability of the cerebellum to inhibit its own climbing fiber input.

7. IMPLICATIONS FOR CEREBELLAR LEARNING FROM THE PHENOMENON OF SAVINGS

The behavioral phenomenon of savings is straightforward—the rate of relearning (after prior experience with acquisition and then extinction) is faster than the rate of original learning. Eyelid conditioning displays robust savings. Whereas the typical animal may require 100–150 training trials to learned robust responses, relearning after extinction can occur in as few as 10–20 trials. This is true even after numerous extinction sessions where no conditioned responses have been seen for more than five sessions (500 trials). This savings is a property of the cerebellum, as it occurs even when mossy fiber stimulation is used as the CS. As such, the important question is, what can this behavioral property of savings tell us about the mechanisms of learning in the cerebellum?

A hypothesis regarding the mechanisms for savings in the cerebellum arose jointly from computer simulation and analytic analysis of cerebellar learning (Medina, Garcia, & Mauk, 2001). The simulations suggested the induction of plasticity at the two sites—initially, there is plasticity at the granule-to-Purkinje synapses in the cerebellar cortex, and then, the associated pauses in Purkinje cells during the CS promoted plasticity at the mossy fiber to nucleus synapses in the deep cerebellar nuclei. During extinction, a similar sequence occurred, the induction of LTP at granule-to-Purkinje synapses removed the pause in Purkinje activity during the CS, and thus "behaviorally," the conditioned responses of the simulation extinguished. From these events, two predictions arise. First, there is a phase during acquisition where Purkinje cells have learned to pause to the CS, but there are no conditioned responses because plasticity in the deep nucleus has not yet developed. Tatsuya Ohyama developed a very clever means to implement an indirect test of this hypothesis. His results support this first prediction (Ohyama & Mauk, 2001). The second prediction was this: for some unknown number of training trials after responses have disappeared during extinction training, the learning-induced strengthening of the mossy fiber synapses in the deep nucleus remains. If retraining is implemented during this phase, savings is observed because the reappearance of conditioned responses requires only the reinduction of LTD in the cerebellar cortex. Moreover, since both simulations (Medina & Mauk, 1999) and analytic analyses (Kenyon, Medina, & Mauk, 1998a, 1998b) had revealed previously that the rate of induction of plasticity in the cerebellar cortex must be faster than the rate of induction of deep nucleus plasticity; otherwise, the system enters perpetual oscillations in synaptic weight changes. Combining these findings led to the hypothesis that (1) the slow induction of plasticity in at the deep nucleus synapses is the rate-limiting factor determining the relatively slow initial acquisition of conditioned responses, (2) learning-induced plasticity at the deep nucleus synapses is only slowly reversed during extinction training, and thus, (3) savings can be explained by this residual plasticity in the deep nucleus as reacquisition of conditioned responses is determined by the rate of plasticity induction in the cerebellar cortex.

With a series of straightforward experiments, Javier Medina set out to test all three of the components of this hypothesis (Medina et al., 2001). He made use of the ability to implement reversible disconnection of the cerebellar cortex from the cerebellar deep nuclei. As discussed above, this unmasks short-latency (i.e., not adaptively timed) conditioned responses that have been shown to be mediated by plasticity in the cerebellar deep nuclei. Thus, the Medina experiments made use of these short-latency responses as a proxy for a direct measure of learning-induced plasticity in the deep nuclei. This approach revealed that short-latency responses are acquired at the same rate as the well-timed responses, consistent with the prediction that the induction of plasticity in the cerebellar nuclei is a rate-limiting factor in the initial acquisition of responses. More amazingly perhaps, these experiments revealed that even in animals that had undergone 40–50 sessions of extinction, and where conditioned responses had not been present for over 40 sessions, short-latency responses can still be unmasked by reversible disconnection of the cerebellar cortex. This observation was strong evidence for the simulation-derived hypothesis that learning-induced plasticity in deep nuclei can remain after extinction. In these experiments, it was found that robust and reliable short-latency responses were seen in animals given 5–15 sessions of extinction training but that there was considerable variability in the status of short-latency responses in animals extinguished for 40–50 sessions. Making use of this variability, these authors found a strong correlation between the magnitudes of short-latency responses at various stages of extinction with the degree of savings seen during reacquisition training the next day. When robust short-latency responses were observed, savings was robust, whereas when short-latency responses were weak or absent, so was savings. These findings offer strong support for the overarching hypothesis that residual plasticity in the cerebellar deep nuclei after extinction training is responsible for savings seen during reacquisition of responses.

What functional significance might this savings mechanism have for cerebellar function under more realistic circumstances? To begin, some forms of cerebellar learning do not show savings—such as adaptation of the vestibuloocular reflex. The difference may be that animals in VOR studies have used this response before the beginning of the experiment, and so, this is not the first time the cerebellum has been engaged to modulate it. In contrast, eyelid conditioning involves about the most unlikely possible combination of stimulus and response one might image: a tone predicts a threat to your eye. Perhaps, this can be a source of insight. The cerebellar cortex contains an enormous number of granule-to-Purkinje synapses, which may make it possible for every combination of stimulus (mossy fiber input) to have the potential to modulate any response. Because there are vastly fewer mossy fiber synapses onto deep nucleus output neurons, it may be necessary to only build synapses between stimuli and responses that need to be modulated. As such, the plasticity induced in the deep nuclei during initial acquisition of conditioned eyelid responses may represent a normal but special case of cerebellar learning that might be characterized as being about resource allocation. The system will only build synapses between mossy fibers that are repeatedly active; at the same time, there is a robust disinhibition of neighboring deep nucleus neurons from the (recently) learned pauses of Purkinje cells. This explanation is consistent with the results from two separate studies showing that there is an increase in the number of excitatory synapses in the cerebellar deep nuclei during eyelid conditioning.

8. SUMMARY

We have outlined briefly a variety of ways that eyelid conditioning studies have contributed to a better understanding of the mechanisms of cerebellar learning. The basis for eyelid conditioning's utility in this regard is the relatively direct way in which it engages the cerebellum. Eyelid conditioning is by no stretch of the imagination a natural behavior that would exemplify the selective pressures that produced the features of the cerebellum. Yet, the artificial nature of eyelid conditioning is both a limitation and a strength. It is not well suited to provide insight as to how the cerebellum may be engaged normally outside of the laboratory. One example of this is a recent study where the procedures of eyelid conditioning had to be altered to better mimic real-world demands in order to study the cerebellar mechanisms that adapt response amplitude (). On the other hand, it is a tremendous advantage of eyelid conditioning that it confers to the experimenter a great deal of control over inputs and a convenient proxy for cerebellar output. Perhaps, this is one contribution that cerebellar research can offer the field of systems neuroscience in general—finding behaviors or tractable techniques that permit the control of inputs to a brain system should be a top priority.

REFERENCES

Andersson, G., Garwicz, M., & Hesslow, G. (1988). Evidence for a GABA-mediated cerebellar inhibition of the inferior olive in the cat. *Experimental Brain Research, 72*(3), 450–456.

Boyden, E. S., Katoh, A., & Raymond, J. L. (2004). Cerebellum-dependent learning: The role of multiple plasticity mechanisms. *Annual Review of Neuroscience, 27*, 581–610.

Buonomano, D. V., & Mauk, M. D. (1994). Neural network model of the cerebellum: Temporal discrimination and the timing of motor responses. *Neural Computation, 6*(1), 38–55.

Cegavske, C. F., Patterson, M. M., & Thompson, R. F. (1979). Neuronal unit activity in the abducens nucleus during classical conditioning of the nictitating membrane response in the rabbit (Oryctolagus cuniculus). *Journal of Comparative Physiology and Psychology, 93*(4), 595–609.

Choi, J. S., & Moore, J. W. (2003). Cerebellar neuronal activity expresses the complex topography of conditioned eyeblink responses. *Behavioral Neuroscience, 117*(6), 1211–1219.

Garcia, K. S., & Mauk, M. D. (1998). Pharmacological analysis of cerebellar contributions to the timing and expression of conditioned eyelid responses. *Neuropharmacology, 37*, 471–480.

Garcia, K. S., Steele, P. M., & Mauk, M. D. (1999). Cerebellar cortex lesions prevent acquisition of conditioned eyelid responses. *Journal of Neuroscience, 19*, 10940–10947.

Gormezano, I. (1972). Investigations of defense and reward conditioning in the rabbit. In A. H. Black, & W. F. Prokasy (Eds.), *Classical conditioning II: Current theory and research* (pp. 151–181). New York: Appleton-Century Crofts.

Gormezano, I., Kehoe, E. J., & Marshall, B. S. (1983). Twenty years of classical conditioning with the rabbit. *Progress in Psychobiology and Physiological Psychology, 10*, 197–275.

Gormezano, I., Prokasy, W. F., & Thompson, R. F. (1987). *Classical conditioning* (3rd ed.). Hillsdale, NJ: Lawrence Earlbaum Associates.

Gormezano, I., Schneiderman, N., Deaux, E., & Fuentes, I. (1962). *Science, 13*, 33–34.

Hilgard, E. R., & Marquis, D. O. (1940). *Conditioning and learning*. New York: Appleton-Century-Crofts.

Ito, M. (1988). Long term depression. *Annual Review of Neuroscience, 12*, 85–102.
Ito, M., & Kano, M. (1982). Long lasting depression of parallel fiber-Purkinje cell transmission induced by conjunctive stimulation of parallel fibers and climbing fibers in the cerebellar cortex. *Neuroscience Letters, 33*, 253–258.
Ito, M., Sakurai, M., & Tongroach, P. (1982). Climbing fibre induced depression of both mossy fibre responsiveness and glutamate sensitivity of cerebellar Purkinje cells. *Journal of Physiology, 324*, 113–134.
Kalmbach, B. E., Davis, T., Ohyama, T., Nores, W. L., & Mauk, M. D. (2010). Cerebellar cortex contributions to the expression and timing of conditioned eyelid responses. *Journal of Neurophysiology, 103*, 2039–2049.
Kalmbach, B. E., Ohyama, T., Kreider, J. C., Riusech, F., & Mauk, M. D. (2009). Cerebellum-forebrain interactions revealed by trace eyelid conditioning. *Learning & Memory, 16*(1), 86–95.
Kalmbach, B. E., Ohyama, T., & Mauk, M. D. (2010). Temporal patterns of inputs to cerebellum necessary and sufficient for trace eyelid conditioning. *Journal of Neurophysiology, 104*(2), 627–640.
Kalmbach, B. E., Voicu, H., Ohyama, T., & Mauk, M. D. (2011). A subtraction mechanism of temporal coding in cerebellum. *Journal of Neuroscience, 31*(6), 2025–2034.
Kehoe, E. J., & Napier, R. M. (1991). In the blink of an eye: Real time stimulus factors in delay and trace conditioning of the rabbit's nictitating membrane response. *Quarterly Journal of Experimental Psychology, 43B*, 257–277.
Kenyon, G. T., Medina, J. F., & Mauk, M. D. (1998a). A mathematical model of the cerebellar-olivary system I: Self-regulating equilibrium of climbing fiber activity. *Journal of Computational Neuroscience, 5*, 17–33.
Kenyon, G. T., Medina, J. F., & Mauk, M. D. (1998b). A mathematical model of the cerebellar-olivary system II. Motor adaptation through systematic disruption of climbing fiber equilibrium. *Journal of Computational Neuroscience, 5*, 71–90.
Lev-Ram, V., Mehta, S. B., Kleinfeld, D., & Tsien, R. Y. (2003). Reversing cerebellar long-term depression. *Proceeding of the National Academy of Sciences of the United States of America, 26*, 15989–15993.
Marr, D. (1969). A theory of cerebellar cortex. *Journal of Physiology, 202*, 437–470.
Mauk, M. D., & Donegan, N. H. (1997). A model of Pavlovian eyelid conditioning based on the synaptic organization of the cerebellum. *Learning & Memory, 3*, 130–158.
Mauk, M. D., & Ruiz, B. P. (1992). The learning-dependent timing of Pavlovian eyelid responses: Differential conditioning using multiple inter-stimulus intervals. *Behavioral Neuroscience, 106*, 666–681.
McCormick, D. A., Lavond, D. G., & Thompson, R. F. (1982). Concomitant classical conditioning of the rabbit nictitating membrane and eyelid responses: Correlations and implications. *Physiology & Behavior, 28*(5), 769–775.
McCormick, D. A., & Thompson, R. F. (1984). Cerebellum: Essential involvement in the classically conditioned eyelid response. *Science, 223*, 296–299.
Medina, J. F., Carey, M. R., & Lisberger, S. G. (2005). The representation of time for motor learning. *Neuron, 45*(1), 157–167.
Medina, J. F., Garcia, K. S., & Mauk, M. D. (2001). A mechanism for savings in the cerebellum. *Journal of Neuroscience, 21*, 4081–4089.
Medina, J. F., Garcia, K. S., Nores, W. L., Taylor, N. M., & Mauk, M. D. (2000). Timing mechanisms in the cerebellum: Testing predictions of a large scale computer simulation. *Journal of Neuroscience, 20*, 5516–5525.
Medina, J. F., & Lisberger, S. G. (2008). Links from complex spikes to local plasticity and motor learning in the cerebellum of awake-behaving monkeys. *Nature Neuroscience, 11*(10), 1185–1192.

Medina, J. F., & Mauk, M. D. (1999). Simulations of cerebellar motor learning: Computational analysis of plasticity at the mossy fiber to deep nucleus synapses. *Journal of Neuroscience, 19*, 7140–7151.

Medina, J. F., & Mauk, M. D. (2000). Computer simulation of cerebellar information processing. *Nature Neuroscience, 3*, 1205–1211.

Medina, J. F., Nores, W. L., & Mauk, M. D. (2002). Inhibition of climbing fibres is a signal for the extinction of conditioned eyelid responses. *Nature, 416*, 330–333.

Miles, F. A., & Lisberger, S. G. (1981). Plasticity in the vestibulo-ocular reflex: A new hypothesis. *Annual Review of Neuroscience, 4*, 273–299.

Millenson, J. R., Kehoe, E. J., & Gormezano, I. (1977). Classical conditioning of the rabbit's nictitating membrane response under fixed and mixed CS-US intervals. *Learning and Motivation, 8*, 351–366.

Ohyama, T., & Mauk, M. D. (2001). Latent learning of timed responses in the cerebellar cortex. *Journal of Neuroscience, 21*, 682–690.

Ohyama, T., Medina, J. F., Nores, W. L., & Mauk, M. D. (2002). Trying to understand the cerebellum well enough to build one. *Annals of the New York Academy of Sciences, 978*, 1–15.

Ohyama, T., Voicu, H., Kalmbach, B. E., & Mauk, M. D. (2010). A decrementing form of plasticity apparent in cerebellar learning. *Journal of Neuroscience, 30*(50), 16993–17003.

Perrett, S. P., & Mauk, M. D. (1995). Extinction of conditioned eyelid responses requires the anterior lobe of cerebellar cortex. *Journal of Neuroscience, 15*(3), 2074–2080.

Perrett, S. P., Ruiz, B. P., & Mauk, M. D. (1993). Cerebellar cortex lesions disrupt the learning-dependent timing of conditioned eyelid responses. *Journal of Neuroscience, 13*(4), 1708–1718.

Sakurai, M. (1987). Synaptic modification of parallel fiber-Purkinje cell transmission in *in vitro* guinea-pig cerebellar slices. *Journal of Physiology, 394*, 463–480.

Schneiderman, N., & Gormezano, I. (1964). Conditioning of the nictitating membrane of the rabbit as a function of the CS-US interval. *Journal of Comparative and Physiological Psychology, 57*, 188–195.

Thompson, R. F. (1976). The search for the engram. *American Psychologist, 31*(3), 209–227.

Thompson, R. F. (2013). An essential memory trace found. *Behavioral Neuroscience, 127*(5), 669–675.

CHAPTER THREE

Cerebellar Long-Term Potentiation: Cellular Mechanisms and Role in Learning

Giorgio Grasselli, Christian Hansel[1]
Department of Neurobiology, University of Chicago, Chicago, Illinois, USA
[1]Corresponding author: e-mail address: chansel@bsd.uchicago.edu

Contents

Abstract

Activity-dependent long-term plasticity of synaptic transmission, such as in long-term potentiation (LTP) and long-term depression (LTD), provides a cellular correlate of experience-driven learning. While at excitatory synapses in the hippocampus and neocortex LTP is seen as the primary learning mechanism, it has been widely assumed that cerebellar motor learning is mediated by LTD at parallel fiber (PF)–Purkinje cell synapses instead. However, recent work on mouse mutants with deficits in AMPA receptor internalization has demonstrated that motor learning can occur in the absence of LTD, suggesting that LTD is not essential. Another recent study has shifted attention toward LTP at PF synapses, showing that blockade of LTP severely affects motor learning. Here, we review the cellular and molecular events that are involved in LTP induction and discuss whether LTP might indeed play a more significant role in cerebellar learning than previously anticipated.

1. THE SEARCH FOR THE MEMORY ENGRAM IN CEREBELLAR LEARNING

In classic Marr–Albus theories of cerebellar learning, synaptic modification plays a central role, remarkably before the discovery of hippocampal LTP by Bliss and Lomo (Bliss & Lomo, 1973). Marr suggested that PF

International Review of Neurobiology, Volume 117
ISSN 0074-7742
http://dx.doi.org/10.1016/B978-0-12-420247-4.00003-8

synapses would be potentiated when activated together with the climbing fiber (CF) input (Marr, 1969). Albus similarly assumed that CF coactivity drives cerebellar learning but predicted a depression at PF synapses as a consequence of paired activity in these two pathways (Albus, 1971). In 1982, it was experimentally demonstrated by Ito and colleagues that indeed prolonged, paired PF and CF activation leads to a lasting depression of PF synapses (Ito & Kano, 1982; Ito, Sakurai, & Tongroach, 1982). Soon after, LTD at PF synapses was widely accepted as the mechanism underlying cerebellar learning, since LTD was expected to remove the inhibition that target neurons in the cerebellar nuclei receive from Purkinje cells, thus facilitating the generation of a cerebellar output signal that may activate the red nucleus. Subsequent studies using genetically modified mice generally confirmed the Marr–Albus–Ito theory by showing that blockade of LTD was reliably correlated with impairment of motor learning (e.g., Aiba et al., 1994; De Zeeuw et al., 1998; Hansel et al., 2006). However, the hypothesis that LTD alone can mediate motor learning has been challenged by two more recent mouse studies. First, it was demonstrated that mice with a blockade of LTP—which can be induced by PF activity alone (Coesmans, Weber, De Zeeuw, & Hansel, 2004; Lev-Ram, Wong, Storm, & Tsien, 2002)—show severe motor learning impairment (Schonewille et al., 2010) suggesting that LTD is not the only type of synaptic plasticity needed for learning. Second, it was shown that mice with a deficit in AMPA receptor endocytosis and LTD display no motor learning impairment at all, which has been interpreted as an indication that LTD is not essential for cerebellar motor learning (Schonewille et al., 2011). Thus, the role of LTD in cerebellar learning needs to be reinvestigated with a focus on its interaction with complementing types of plasticity. Here, we discuss LTP at PF synapses and its validity as a novel correlate of learning.

2. LTP: CELLULAR MECHANISMS

PF-LTP is induced by isolated PF activation at 1 Hz for 5 min (Coesmans et al., 2004; Lev-Ram et al., 2002). Three lines of evidence suggest that this form of LTP is postsynaptically expressed: (a) It is not accompanied by changes in the paired-pulse facilitation ratio (Coesmans et al., 2004; Lev-Ram et al., 2002); (b) the low-affinity competitive AMPA receptor antagonist γ-D-glutamylglycine, which can be used as a reporter for changes in glutamate release (Wadiche & Jahr, 2001), does not block PF-EPSCs to a lower degree after LTP induction than before (this effect would be expected if glutamate release was upregulated; Coesmans et al., 2004); and (c) LTP can reverse postsynaptic LTD (Coesmans et al., 2004; Lev-Ram, Mehta, Kleinfeld, & Tsien, 2003). Remarkably, both LTP and LTD can result from PF activation at 1 Hz for 5 min, but the latter requires coactivation of the CF input, which leads to supralinear calcium transients in PF-contacted spines (Wang, Denk, & Häusser, 2000). This larger calcium signal promotes the induction of LTD, which has a higher calcium threshold than LTP (Coesmans et al., 2004). LTD and LTP can also be induced using 100 Hz PF burst stimulation (PF bursts evoked by eight stimuli at 100 Hz are repetitively applied at 1 Hz for 5 min), and at this higher stimulus frequency, the LTD-dependence on CF coactivation is maintained (Piochon, Levenes, Ohtsuki, & Hansel, 2010). LTD induction requires large calcium transients (Coesmans et al., 2004), which partially result from the activation of type 1 metabotropic glutamate receptors (mGluR1; Aiba et al., 1994) and subsequent calcium release from internal stores (Finch & Augustine, 1998; Takechi, Eilers, & Konnerth, 1998) as well as—in the mature brain—from activation of NMDA receptors (Piochon et al., 2010), which are postsynaptically expressed at CF-Purkinje cell synapses (He, Titley, Grasselli, Piochon, & Hansel, 2013; Piochon et al., 2007; Renzi, Farrant, & Cull-Candy, 2007). In addition, LTD induction depends on the activation of protein kinase C (PKC; De Zeeuw et al., 1998; Linden & Connor, 1991) and α-calcium-/calmodulin-dependent protein kinase II (Hansel et al., 2006). While PKC-α phosphorylates the AMPA receptor subunit GluA2 and initiates AMPA receptor endocytosis (Chung, Steinberg, Huganir, & Linden, 2003; Leitges, Kovac, Plomann, & Linden, 2004), CaMKII negatively regulates phosphodiesterase 1, thus facilitating a cGMP-/PKG-mediated suppression of protein phosphatase 2A (PP2A; Kawaguchi & Hirano, 2013).

In contrast to the kinase-dependent LTD induction process, LTP induction depends on the activation of protein phosphatases, thus providing a mirror image of LTP/LTD induction requirements at hippocampal synapses (Jörntell & Hansel, 2006). LTP is prevented when the calcium chelator BAPTA is added to the internal saline at high concentrations (30 mM; Coesmans et al., 2004) suggesting that the induction process depends on calcium ions at resting or moderately elevated levels. LTP induction is blocked in the presence of okadaic acid and microcystin-LR (Belmeguenai & Hansel, 2005), which are inhibitors of protein phosphatases 1 (PP1) and PP2A, respectively. LTP is also blocked in the presence of the PP1 inhibitor peptide inhibitor-2, the PP2A inhibitor fostriecin, and the PP2B (calcineurin) inhibitor cyclosporin A, respectively (Fig. 3.1;

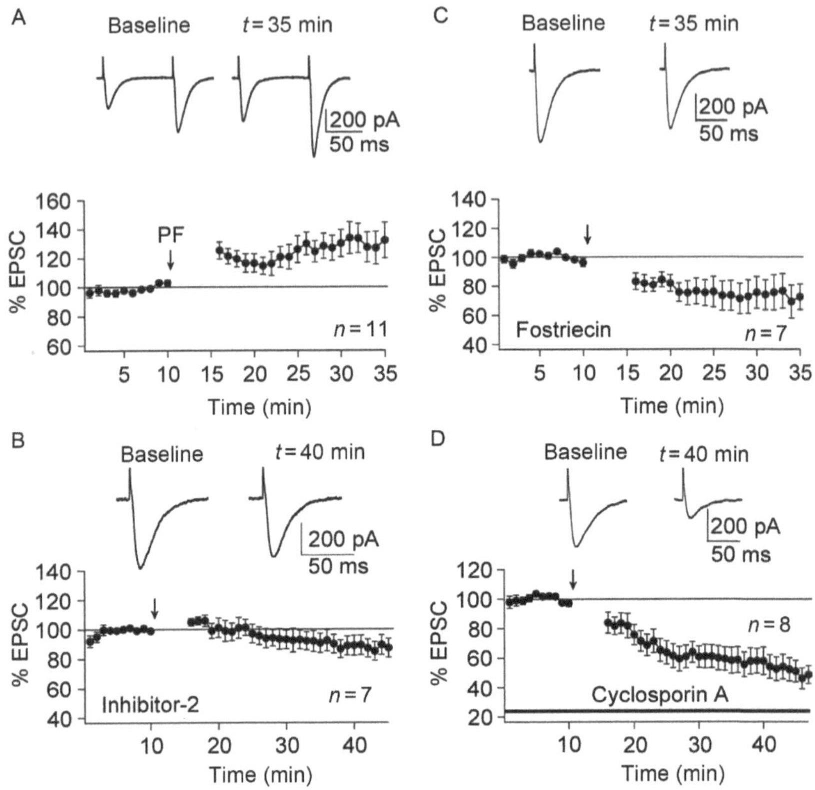

Figure 3.1 PP1, PP2A, and PP2B are required for PF-LTP induction. (A) PF-LTP is induced by PF stimulation at 1 Hz for 5 min ($n=11$). Each data point represents the average of three successive responses evoked at 0.05 Hz. The top traces show EPSCs before and after the application of the LTP protocol (arrow). (B) Inhibitor-2, a PP1 inhibitory protein, blocks PF-LTP induction when added to the pipette saline (100 nM; $n=7$). (C) The PP2A inhibitor fostriecin, which is also added to the pipette saline (50 nM), blocks LTP induction ($n=7$). (D) Bath application of the PP2B-selective inhibitor cyclosporin A (100 μM) blocks LTP induction as well ($n=8$). The bar indicates the presence of cyclosporin A in the bath. Error bars are mean ± SEM. *This figure is modified from Belmeguenai and Hansel (2005). Copyright 2005 by the Society for Neuroscience.*

Belmeguenai & Hansel, 2005), suggesting that all three protein phosphatases present in neurons, PP1, PP2A, and calcineurin, are involved in LTP induction. The critical role of calcineurin was confirmed in mice with a Purkinje cell-specific knockout of calcineurin (L7-PP2B mice), in which LTP is absent, but LTD is unaffected (Fig. 3.2; Schonewille et al., 2010). Finally, adding active calcineurin to the pipette saline prevents LTD induction but has no effect on LTP (Belmeguenai & Hansel, 2005). The need for

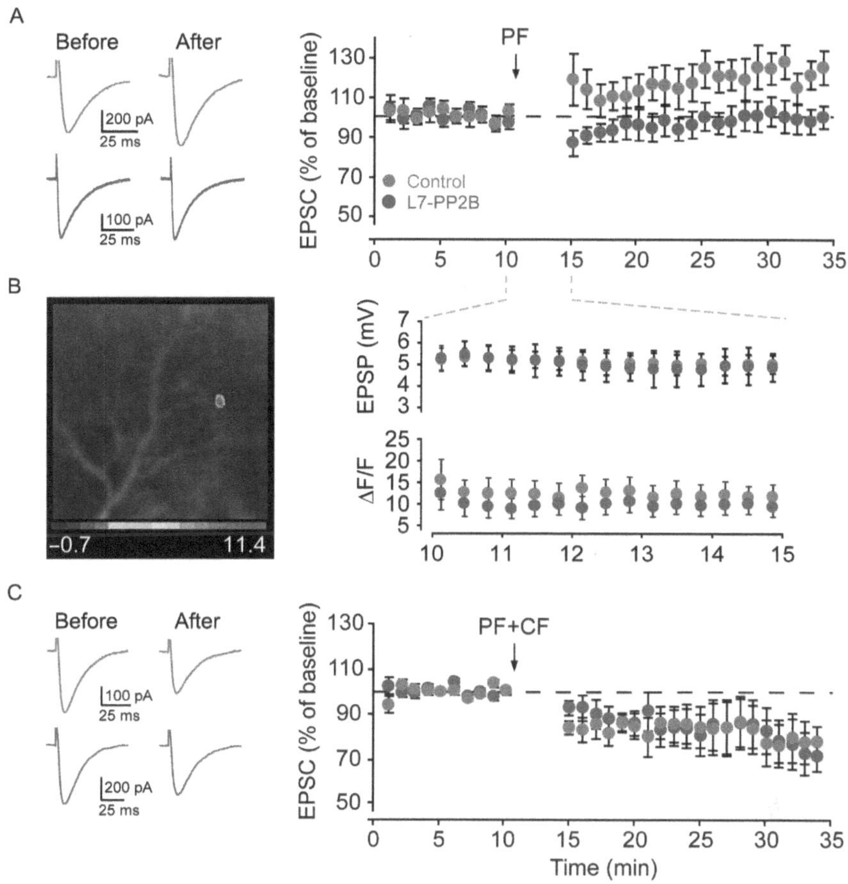

Figure 3.2 Mice with a Purkinje cell-specific knockout of calcineurin (L7-PP2B) show impaired PF-LTP. (A) PF-LTP induction was impaired in slices obtained from L7-PP2B mice ($n=8$) but was intact in the slices prepared from wild-type mice ($n=7$). (B) Right, EPSPs and calcium transient peaks during the tetanus did not differ between L7-PP2B ($n=10$) and WT mice ($n=13$). Left, sample image of a PF stimulation-induced calcium signal. Middle, example trace of PF stimulation-elicited calcium transients (average of three). (C) PF-LTD did not differ between L7-PP2B ($n=7$) and WT mice ($n=9$), respectively. PF-LTP was induced by PF stimulation at 1 Hz for 5 min, while LTD was induced by paired PF and CF stimulation at 1 Hz for 5 min. Traces on the left show EPSCs before (left) and after (right) the application of the tetanization protocol. Error bars are mean ± SEM. *This figure is taken from Schonewille et al. (2010). Copyright 2010 by Elsevier.*

activation of calcineurin, the only calcium-/calmodulin-sensitive phosphatase of these three phosphatases, suggests that LTP indeed requires a surge in calcium levels. It is currently not known how PP1, PP2A, and calcineurin act to promote LTP induction, but it is conceivable that a kinase/phosphatase switch exists, similar to the inversely operating bimodal CaMKII/PP1 switch that has been suggested for hippocampal synapses, in which calcineurin activity prevents inhibition of PP1, which directly counteracts CaMKII (Lisman, 1989; Lisman & Zhabotinsky, 2001). In contrast to LTD, LTP induction does not require the activation of mGluR1 receptors (Fig. 3.3; Belmeguenai et al., 2008) or NMDA receptors (Piochon et al., 2010), leaving voltage-gated calcium channels as a calcium source.

In the initial description of postsynaptic PF-LTP, it was shown that LTP depends on nitric oxide (NO) signaling (Lev-Ram et al., 2002). LTP is

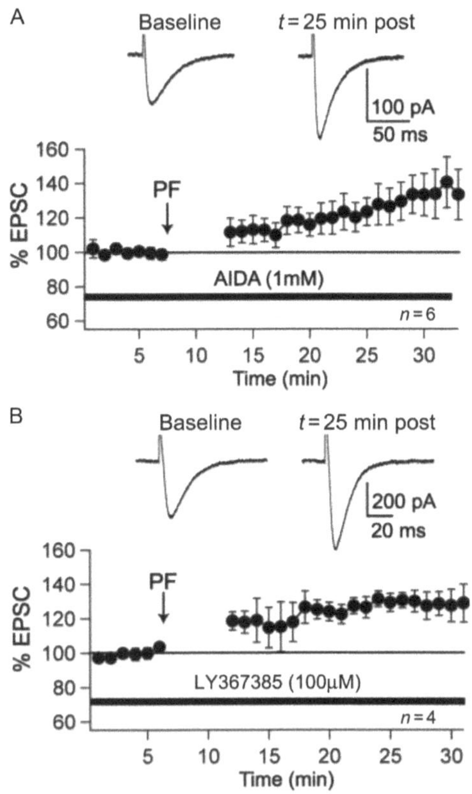

Figure 3.3 PF-LTP induction does not depend on mGluR1 activation. (A) LTP can be induced when the group I mGluR antagonist AIDA (1 mM) is present in the bath ($n=6$). (B) LTP is also induced in the presence of the mGluR1a antagonist LY367385 (100 μM) in the bath ($n=4$). In (A) and (B), traces show PF-EPSCs before and 25 min after tetanization (arrows). The bars indicate the presence of AIDA and LY367385, respectively, in the bath. Error bars are mean ± SEM. *This figure is taken from Belmeguenai et al. (2008). Copyright 2008 by the American Physiological Society.*

mediated by GluA2 membrane insertion after binding of GluA2 to N-ethylmaleimide-sensitive factor (NSF; Kakegawa & Yuzaki, 2005), a process that is triggered by NO through S-nitrosylation of NSF (Huang et al., 2005). Thus, in cerebellar LTP, GluA2 membrane insertion is not just a constitutive delivery mechanism as previously suggested (Shi, Hayashi, Esteban, & Malinow, 2001), but the NO dependence of this signaling step indicates that it can be activity-driven instead (see Hansel, 2005).

3. CONFLICTING DATA ON THE ROLE OF CALCINEURIN IN LTD INDUCTION

Our model of bidirectional plasticity at PF–Purkinje cell synapses suggests that LTD depends on a kinase signaling pathway, while LTP requires the activation of protein phosphatases PP1, PP2A, and calcineurin (see Jörntell & Hansel, 2006). The implication that calcineurin promotes LTP, but not LTD induction, has recently been challenged. One group has reported that LTD can be blocked by the calcineurin inhibitors cyclosporin A and FK506, respectively, or by adding a calcineurin autoinhibitory peptide to the pipette saline (Fujiwara, Kakizawa, & Iino, 2007). Another group similarly demonstrated blockade of LTD when the calcineurin autoinhibitory peptide was added to the pipette saline (Nomura et al., 2012). How can these conflicting data be reconciled? First of all, the assumption that LTD relies on a kinase-driven pathway, while LTP relies on a phosphatase-driven pathway does not entirely exclude the possibility that calcineurin could play some supportive role in LTD, such as a regulation of the auxiliary protein stargazin (Nomura et al., 2012). The use of different LTD induction protocols might lead to different sensitivities to interference with such processes. In this context, it is important to note that Fujiwara et al. (2007) and Nomura et al. (2012) both used LTD protocols, in which PF stimulation is paired with Purkinje cell depolarization, whereas we use the more physiological PF + CF coactivation in our studies (Belmeguenai & Hansel, 2005; Schonewille et al., 2010). It also needs to be pointed out that the application of cyclosporin A and FK506, respectively, reduced LTD, but did not prevent it, still allowing for LTD amplitudes of ~15% (Fujiwara et al., 2007). Finally, the application of inhibitory peptides through the patch pipette (Fujiwara et al., 2007; Nomura et al., 2012) can be problematic, because there is no real control baseline in these recordings (the peptide diffuses into the cell as soon as the whole-cell patch configuration is established) and because it is difficult to ascertain specificity of the observed effects when adding potentially sticky external peptides to the soma, which are expected to act locally at the synapses.

We maintain our claim that calcineurin is required for LTP, but not for LTD induction, based on the following demonstration:

a. Application of the LTP protocol in the presence of cyclosporin A does not cause LTP, but pronounced LTD instead (Fig. 3.1D; Belmeguenai & Hansel, 2005).

b. When active calcineurin is added to the pipette saline, the application of the LTD protocol (PF + CF) causes mild potentiation instead (Belmeguenai & Hansel, 2005). Note that neither active calcineurin nor any of the drugs mentioned in (a) causes changes in synaptic strength in the absence of tetanization (Belmeguenai & Hansel, 2005).

c. In mice with a Purkinje cell-specific knockout of calcineurin (L7-PP2B), LTP is blocked, but LTD remains unaffected (Fig. 3.2C; Schonewille et al., 2010).

Finally, several studies from other laboratories have demonstrated that LTD induction requires, or is supported by, the suppression of PP1/2A (Ajima & Ito, 1995; Eto, Bock, Brautigan, & Linden, 2002; Kawaguchi & Hirano, 2013; Launey, Endo, Sakai, Harano, & Ito, 2004) and that a late phase of LTD requires a downregulation of calcineurin (Fuji & Hirano, 2002). Together, these studies support the notion that phosphatases PP1, PP2A, and PP2B negatively regulate LTD rather than promote it (also note that calcineurin and PP1 have been suggested to act in tandem; Lisman, 1989).

4. LTP AND CEREBELLAR LEARNING

The recent demonstration that mutant mice with a deficit in LTD—resulting from a constitutive blockade of AMPA receptor endocytosis in a non-tissue-specific manner—show intact motor learning (eyeblink conditioning and gain adaptation of the vestibuloocular reflex, VOR) has motivated the claim that LTD is not essential for cerebellar motor learning (Schonewille et al., 2011). A study published a year earlier has shown that, in contrast, both forms of motor learning are severely impaired in mice with a Purkinje cell-specific knockout of calcineurin, which leads to a blockade of synaptic and nonsynaptic potentiation mechanisms (LTP and intrinsic plasticity, respectively; Schonewille et al., 2010). Thus, these two studies together may seem to suggest that LTP, but not LTD, is the cellular correlate of cerebellar motor learning. In this context, it is of interest to note that fear conditioning, a paradigm of non-motor learning that involves the cerebellum, has also been shown to be associated with a potentiation, but not a depression, at PF synapses (Sacchetti, Scelfo, Tempia, & Strata, 2004; Zhu, Scelfo, Hartell, Strata, & Sacchetti, 2007).

However, we believe that cerebellar motor learning is mediated neither by LTD nor by LTP alone but that both forms of PF plasticity play an important and complementary role. We currently view motor learning as a process involving multiple learning sites and forms of plasticity (see also Hansel, Linden, & D'Angelo, 2001). Among these, two plasticity sites have been shown to be particularly important for learning: first, LTP also occurs at mossy fiber synapses onto cerebellar nuclei neurons, establishing a learning site outside of the cerebellar cortex (Medina & Mauk, 1999; Ohyama, Nores, Medina, Riusech, & Mauk, 2006; Pugh & Raman, 2006, 2009). Second, the PF–Purkinje cell synapse provides a learning site within the cerebellar cortex that permits regulation of plasticity in the nuclei, but plasticity at this cortical site results from more complex processing of a richer set of sensory input/feedback that is received by Purkinje cell dendrites. LTD and LTP are two sides of the same coin and should not be viewed independently in learning models, because one will act as the learning mechanism and the other as the reversal mechanism. In VOR gain conditioning, LTD and LTP are indeed considered to play an opposite and complementary role in a reversible way, underlying an increase (gain-up) and a decrease (gain-down) of gain, respectively (Boyden & Raymond, 2003). For example, an acute pharmacological block of mGluR1 receptors, which are necessary for the induction of LTD, prevents an increase of gain, but does not affect the other side of this type of learning; activity-dependent gain decreases (Titley, Heskin-Sweezie, & Broussard, 2010). We argue that a chronic blockade of the reversal mechanism (e.g., genetic manipulation in mice) would saturate the learning mechanism, possibly leading to the same behavioral phenotype as a blockade of the learning mechanism itself.

If PF-LTD is one of several important types of plasticity involved in cerebellar motor learning, why can motor learning be intact when LTD is genetically impaired (Schonewille et al., 2011)? The assumption of two learning sites—in the cerebellar cortex and in the nuclei—allows for the possibility that, under some conditions, plasticity at one of these sites, here in the cerebellar nuclei, can be sufficient to mediate learning. We believe that this is the most likely explanation for the intact motor learning observed in PF-LTD-deficient mice (Schonewille et al., 2011). It has to be kept in mind that the two types of motor learning paradigms that are routinely used in laboratories—eyeblink conditioning and VOR gain conditioning—are very simple types of motor learning, because rodents that are raised under conditions of extreme sensory deprivation (as compared to their natural environment) are in these learning tests exposed to a monotone, repetitive application of the same stimulus. It seems conceivable that a learned response to this type of reduced input can be produced when parts of the otherwise more complex learning machinery are missing.

In summary, we favor the view that complex motor learning requires LTD/LTP at PF synapses as well as plasticity in the nuclei. Bidirectional PF plasticity is complemented and regulated by additional types of plasticity, for example, LTD of the instructive CF "teacher signal" (Hansel & Linden, 2000) as well as Purkinje cell intrinsic plasticity (Belmeguenai et al., 2010; Ohtsuki, Piochon, Adelman, & Hansel, 2012). Postsynaptically expressed forms of PF-LTD and LTP can reverse each other (Coesmans et al., 2004) allowing for fast adaptation to environmental needs. It is interesting to note that a presynaptic form of PF-LTP, which can be induced by 8 Hz PF tetanization for 15 s (Salin, Malenka, & Nicoll, 1996), is blocked by a process involving retrograde cannabinoid signaling and activation of CB1 receptors on PF terminals under "LTD conditions" such as CF coactivation (Van Beugen, Nagaraja, & Hansel, 2006) or cholinergic signaling (Rinaldo & Hansel, 2013). This suppression of presynaptic LTP may promote the functional penetrance of LTD.

ACKNOWLEDGMENTS

Our work is supported by a grant from the National Institute of Neurological Disorders and Stroke (NS-062271 to C. H.). We would like to thank the members of the Hansel laboratory for their helpful discussions.

REFERENCES

Aiba, A., Kano, M., Chen, C., Stanton, M. E., Fox, G. D., Herrup, K., et al. (1994). Deficient cerebellar long-term depression and impaired motor learning in mGluR1 mutant mice. *Cell*, *79*, 377–388.

Ajima, A., & Ito, M. (1995). A unique role of protein phosphatases in cerebellar long-term depression. *NeuroReport*, *6*, 297–300.

Albus, J. S. (1971). A theory of cerebellar function. *Mathematical Biosciences*, *10*, 25–61.

Belmeguenai, A., Botta, P., Weber, J. T., Carta, M., De Ruiter, M., De Zeeuw, C. I., et al. (2008). Alcohol impairs long-term depression at the cerebellar parallel fiber-Purkinje cell synapse. *Journal of Neurophysiology*, *100*, 3167–3174.

Belmeguenai, A., & Hansel, C. (2005). A role for protein phosphatases 1, 2A, and 2B in cerebellar long-term potentiation. *Journal of Neuroscience*, *25*, 10768–10772.

Belmeguenai, A., Hosy, E., Bengtsson, F., Pedroarena, C. M., Piochon, C., Teuling, E., et al. (2010). Intrinsic plasticity complements long-term potentiation in parallel fiber input gain control in cerebellar Purkinje cells. *Journal of Neuroscience*, *30*, 13630–13643.

Bliss, T. V. P., & Lomo, T. (1973). Long-lasting potentiation of synaptic transmission in the dentate area of the anesthetized rabbit following stimulation of the perforant path. *Journal of Physiology*, *232*, 331–356.

Boyden, E. S., & Raymond, J. L. (2003). Active reversal of motor memories reveals rules governing memory encoding. *Neuron*, *39*, 1031–1042.

Chung, H. J., Steinberg, J. P., Huganir, R. L., & Linden, D. J. (2003). Requirement of AMPA receptor GluR2 phosphorylation for cerebellar long-term depression. *Science*, *300*, 1751–1755.

Coesmans, M., Weber, J. T., De Zeeuw, C. I., & Hansel, C. (2004). Bidirectional parallel fiber plasticity in the cerebellum under climbing fiber control. *Neuron*, *44*, 691–700.

De Zeeuw, C. I., Hansel, C., Bian, F., Koekkoek, S. K., van Alphen, A. M., Linden, D. J., et al. (1998). Expression of a protein kinase C inhibitor in Purkinje cells blocks cerebellar LTD and adaptation of the vestibulo-ocular reflex. *Neuron*, *20*, 495–508.

Eto, M., Bock, R., Brautigan, D. L., & Linden, D. J. (2002). Cerebellar long-term synaptic depression requires PKC-mediated activation of CPI-17, a myosin/moesin phosphatase inhibitor. *Neuron*, *36*, 1145–1158.

Finch, E. A., & Augustine, G. J. (1998). Local calcium signaling by inositol-1,4,5-trisphosphate in Purkinje cell dendrites. *Nature*, *396*, 753–756.

Fuji, H., & Hirano, T. (2002). Calcineurin regulates induction of late phase of cerebellar long-term depression in rat cultured Purkinje neurons. *European Journal of Neuroscience*, *16*, 1777–1788.

Fujiwara, A., Kakizawa, S., & Iino, M. (2007). Induction of cerebellar long-term depression requires activation of calcineurin in Purkinje cells. *Neuropharmacology*, *52*, 1663–1670.

Hansel, C. (2005). When the B-team runs plasticity: GluR2 receptor trafficking in cerebellar long-term potentiation. *Proceedings of the National Academy of Sciences of the United States of America*, *102*, 18245–18246.

Hansel, C., de Jeu, M., Belmeguenai, A., Houtman, S. H., Buitendijk, G. H., Andreev, D., et al. (2006). αCaMKII is essential for cerebellar LTD and motor learning. *Neuron*, *51*, 835–843.

Hansel, C., & Linden, D. J. (2000). Long-term depression of the cerebellar climbing fiber-Purkinje neuron synapse. *Neuron*, *26*, 473–482.

Hansel, C., Linden, D. J., & D'Angelo, E. (2001). Beyond parallel fiber LTD: The diversity of synaptic and non-synaptic plasticity in the cerebellum. *Nature Neuroscience*, *4*, 467–475.

He, Q., Titley, H., Grasselli, G., Piochon, C., & Hansel, C. (2013). Ethanol affects NMDA receptor signaling at climbing fiber-Purkinje cell synapses in mice and impairs cerebellar LTD. *Journal of Neurophysiology*, *109*, 1333–1342.

Huang, Y., Man, H. Y., Sekine-Aizawa, Y., Han, Y., Juluri, K., Luo, H., et al. (2005). S-nitrosylation of N-ethylmaleimide sensitive factor mediates surface expression of AMPA receptors. *Neuron*, *46*, 533–540.

Ito, M., & Kano, M. (1982). Long-lasting depression of parallel fiber-Purkinje cell transmission induced by conjunctive stimulation of parallel fibers and climbing fibers in the cerebellar cortex. *Neuroscience Letters*, *33*, 253–258.

Ito, M., Sakurai, M., & Tongroach, P. (1982). Climbing fibre induced depression of both mossy fibre responsiveness and glutamate sensitivity of cerebellar Purkinje cells. *Journal of Physiology*, *324*, 113–134.

Jörntell, H., & Hansel, C. (2006). Synaptic memories upside down: Bidirectional plasticity at cerebellar parallel fiber-Purkinje cell synapses. *Neuron*, *52*, 227–238.

Kakegawa, W., & Yuzaki, M. (2005). A mechanism underlying AMPA receptor trafficking during cerebellar long-term potentiation. *Proceedings of the National Academy of Sciences of the United States of America*, *102*, 17846–17851.

Kawaguchi, S., & Hirano, T. (2013). Gating of long-term depression by Ca2+/calmodulin-dependent protein kinase II through enhanced cGMP signaling in cerebellar Purkinje cells. *Journal of Physiology, 591*, 1707–1730.

Launey, T., Endo, S., Sakai, R., Harano, J., & Ito, M. (2004). Protein phosphatase 2A inhibition induces cerebellar long-term depression and declustering of synaptic AMPA receptor. *Proceedings of the National Academy of Sciences of the United States of America, 101*, 676–681.

Leitges, M., Kovac, J., Plomann, M., & Linden, D. J. (2004). A unique PDZ ligand in PKCα confers induction of cerebellar long-term synaptic depression. *Neuron, 44*, 585–594.

Lev-Ram, V., Mehta, S. B., Kleinfeld, D., & Tsien, R. Y. (2003). Reversing cerebellar long-term depression. *Proceedings of the National Academy of Sciences of the United States of America, 100*, 15989–15993.

Lev-Ram, V., Wong, S. T., Storm, D. R., & Tsien, R. Y. (2002). A new form of cerebellar long-term potentiation is postsynaptic and depends on nitric oxide but not cAMP. *Proceedings of the National Academy of Sciences of the United States of America, 99*, 8389–8393.

Linden, D. J., & Connor, J. A. (1991). Participation of postsynaptic PKC in cerebellar long-term depression in culture. *Science, 254*, 1656–1659.

Lisman, J. (1989). A mechanism for the Hebb and the anti-Hebb processes underlying learning and memory. *Proceedings of the National Academy of Sciences of the United States of America, 86*, 9574–9578.

Lisman, J., & Zhabotinsky, A. M. (2001). A model of synaptic memory: A CaMKII/PP1 switch that potentiates transmission by organizing an AMPA receptor anchoring assembly. *Neuron, 31*, 191–201.

Marr, D. (1969). Theory of cerebellar cortex. *Journal of Physiology, 202*, 437–455.

Medina, J. F., & Mauk, M. D. (1999). Simulations of cerebellar motor learning: Computational analysis of plasticity at the mossy fiber to deep nucleus synapse. *Journal of Neuroscience, 19*, 7140–7151.

Nomura, T., Kakegawa, W., Matsuda, S., Kohda, K., Nishiyama, J., Takahashi, T., et al. (2012). Cerebellar long-term depression requires dephosphorylation of TARP in Purkinje cells. *European Journal of Neuroscience, 35*, 402–410.

Ohtsuki, G., Piochon, C., Adelman, J. P., & Hansel, C. (2012). SK2 channel modulation contributes to compartment-specific dendritic plasticity in cerebellar Purkinje cells. *Neuron, 75*, 108–120.

Ohyama, T., Nores, W. L., Medina, J. F., Riusech, F. A., & Mauk, M. D. (2006). Learning-induced plasticity in deep cerebellar nucleus. *Journal of Neuroscience, 26*, 12656–12663.

Piochon, C., Irinopoulou, T., Brusciano, D., Bailly, Y., Mariani, J., & Levenes, C. (2007). NMDA receptor contribution to the climbing fiber response in the adult mouse Purkinje cell. *Journal of Neuroscience, 27*, 10797–10809.

Piochon, C., Levenes, C., Ohtsuki, G., & Hansel, C. (2010). Purkinje cell NMDA receptors assume a key role in synaptic gain control in the mature cerebellum. *Journal of Neuroscience, 30*, 15330–15335.

Pugh, J. R., & Raman, I. M. (2006). Potentiation of mossy fiber EPSCs in the cerebellar nuclei by NMDA receptor activation followed by postinhibitory rebound current. *Neuron, 51*, 113–123.

Pugh, J. R., & Raman, I. M. (2009). Nothing can be coincidence: Synaptic inhibition and plasticity in the cerebellar nuclei. *Trends in Neurosciences, 32*, 170–177.

Renzi, M., Farrant, M., & Cull-Candy, S. G. (2007). Climbing-fibre activation of NMDA receptors in Purkinje cells of adult mice. *Journal of Physiology, 585*, 91–101.

Rinaldo, L., & Hansel, C. (2013). Muscarinic acetylcholine receptor activation blocks long-term potentiation at cerebellar parallel fiber-Purkinje cell synapses via cannabinoid signaling. *Proceedings of the National Academy of Sciences of the United States of America, 110*, 11181–11186.

Sacchetti, B., Scelfo, B., Tempia, F., & Strata, P. (2004). Long-term synaptic changes induced in the cerebellar cortex by fear conditioning. *Neuron, 42*, 973–982.
Salin, P. A., Malenka, R. C., & Nicoll, R. A. (1996). Cyclic AMP mediates a presynaptic form of LTP at cerebellar parallel fiber synapses. *Neuron, 16*, 797–803.
Schonewille, M., Belmeguenai, A., Koekkoek, S. K., Houtman, S. H., Boele, H. J., van Beugen, B. J., et al. (2010). Purkinje cell-specific knockout of the protein phosphatase PP2B impairs potentiation and cerebellar motor learning. *Neuron, 67*, 618–628.
Schonewille, M., Gao, Z., Boele, H. J., Vinueza Veloz, M. F., Amerika, W. E., Simek, A. A., et al. (2011). Reevaluating the role of LTD in cerebellar motor learning. *Neuron, 70*, 43–50.
Shi, S., Hayashi, Y., Esteban, J. A., & Malinow, R. (2001). Subunit-specific rules governing AMPA receptor trafficking to synapses in hippocampal pyramidal neurons. *Cell, 105*, 331–343.
Takechi, H., Eilers, J., & Konnerth, A. (1998). A new class of synaptic response involving calcium release in dendritic spines. *Nature, 396*, 757–760.
Titley, H. K., Heskin-Sweezie, R., & Broussard, D. M. (2010). The bidirectionality of motor learning in the vestibulo-ocular reflex is a function of cerebellar mGluR1 receptors. *Journal of Neurophysiology, 104*, 3657–3666.
Van Beugen, B. J., Nagaraja, R. Y., & Hansel, C. (2006). Climbing fiber-evoked endocannabinoid signaling heterosynaptically suppresses presynaptic cerebellar long-term potentiation. *Journal of Neuroscience, 26*, 8289–8294.
Wadiche, J. I., & Jahr, C. E. (2001). Multivesicular release at climbing fiber-Purkinje cell synapses. *Neuron, 32*, 301–313.
Wang, S. S. H., Denk, W., & Häusser, M. (2000). Coincidence detection in single dendritic spines mediated by calcium release. *Nature Neuroscience, 3*, 1266–1273.
Zhu, L., Scelfo, B., Hartell, N. A., Strata, P., & Sacchetti, B. (2007). The effects of fear conditioning on cerebellar LTP and LTD. *European Journal of Neuroscience, 26*, 219–227.

CHAPTER FOUR

The Ontogeny of Associative Cerebellar Learning

John H. Freeman
Department of Psychology and Neuroscience Program, The University of Iowa, Iowa City, Iowa, USA

Contents

Abstract

Ontogenetic changes in associative cerebellar learning have been examined extensively using eyeblink conditioning in infant humans and rats. The cerebellum is essential for eyeblink conditioning in adult and infant animals. The cerebellum receives input from the conditional stimulus (CS) through the pontine mossy fiber projection and unconditional stimulus (US) input through the inferior olive climbing fiber projection. Coactivation of the CS and US pathways induces synaptic plasticity in the cerebellum, which is necessary for the conditional response. Ontogenetic changes in eyeblink conditioning are driven by developmental changes in the projections of subcortical sensory nuclei to the pontine nuclei and in the inhibitory projection from the cerebellar deep nuclei to the inferior olive. Developmental changes in the CS and US pathways limit the induction of learning-related plasticity in the cerebellum and thereby limit acquisition of eyeblink conditioning.

International Review of Neurobiology, Volume 117
ISSN 0074-7742
http://dx.doi.org/10.1016/B978-0-12-420247-4.00004-X

1. INTRODUCTION

The cerebellum is a relatively late developing brain region in mammalian species with altricial infants such as humans, dogs, cats, and rats. For example, in rats, most of the granule cell neurogenesis occurs after birth and there are massive developmental changes in Purkinje cell dendrites and synaptic connectivity during the first several postnatal weeks (Altman, 1972a, 1972b, 1972c, 1975; Altman & Winfree, 1977). The immature state of the infant cerebellum corresponds to immature motor coordination and motor learning in altricial infants. It is therefore reasonable to assume that the development of coordinated movement and motor learning are due to developmental changes in cerebellar plasticity mechanisms. Our research, however, shows that the development of motor learning is driven by developmental changes in cerebellar interactions with its primary input pathways and not by developmental changes in plasticity mechanisms.

We have examined the development of cerebellar function in relation to associative motor learning. Associative motor learning is a type of learning in which a stimulus or cue predicts that a particular movement will be needed. The following review summarizes our research on the neural mechanisms underlying the ontogeny of associative cerebellar learning.

2. BEHAVIORAL ONTOGENY OF EYEBLINK CONDITIONING

The ontogeny of associative cerebellar learning has been examined using eyeblink conditioning in infant humans, rabbits, and rats (Brown & Woodruff-Pak, 2012; Herbert, Eckerman, & Stanton, 2003; Stanton, Freeman, & Skelton, 1992). Eyeblink conditioning is an associative learning procedure that includes the sequential presentation of two stimuli: a conditional stimulus (CS) that does not cause eyelid movement before training such as a tone or light and an unconditional stimulus (US) that elicits eyelid closure before training (the unconditional response, UR) such as a puff of air to the cornea or a brief electrical stimulus near the eye (Fig. 4.1). When the CS is presented prior to the US repeatedly, the organism will begin to produce eyelid closure, nictitating membrane movement, and eyeball retraction during the CS but before the US (Deaux & Gormezano, 1963; Gormezano, Schneiderman, Deaux, & Fuentes, 1962; Schneiderman,

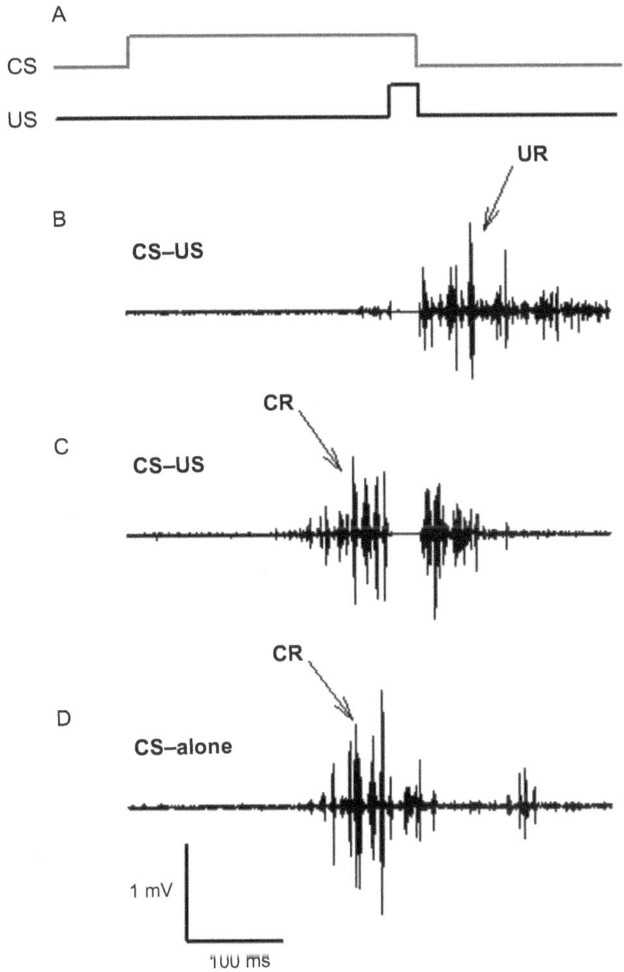

Figure 4.1 Delay eyeblink conditioning. (A) A conditional stimulus (CS) such as a tone or light is presented prior to an unconditional stimulus (US) that elicits an eyelid closure unconditional response (UR). After repeated paired presentations of the CS and US, a conditional eyelid closure response (CR) is elicited before the onset of the US. Eyelid electromyography (EMG) activity is displayed for the first CS–US trial (B), a CS–US trial after learning (C), and during a CS–alone test trial after learning (D). Greater EMG corresponds to more downward eyelid movement.

Fuentes, & Gormezano, 1962). These anticipatory movements constitute the learned or conditional response (CR). This review will focus on the eyelid movement CR, but it is important to note that many of the studies cited here recorded the nictitating membrane CR. Maximum eyelid closure during the CR is precisely timed to the onset time of the US. The eyeblink CR therefore provides measures of predictive learning (i.e., the CS predicts that the US is coming) and learned response timing. Eyeblink conditioning depends upon the cerebellum and at least two sites of synaptic plasticity (see discussion below).

Learning in eyeblink conditioning is typically measured as an increase in the incidence and amplitude of the CR across training trials (i.e., each CS–US presentation). Adult organisms require 100–200 trials to show maximum levels of conditioning. Infants have a slower rate of increase in CRs across training trials and reach a lower maximum level of conditioning (Stanton et al., 1992). The rate and magnitude of eyeblink conditioning increase as a nearly linear function of postnatal age between postnatal days 17 and 31. The nearly linear increase in eyeblink conditioning suggests that maturational processes occur during postnatal development that support progressively stronger cerebellar learning.

3. NEURAL MECHANISMS UNDERLYING EYEBLINK CONDITIONING

The following section reviewing the neural mechanisms underlying eyeblink conditioning is limited to findings that are directly relevant to the ontogeny of eyeblink conditioning. A more comprehensive review of this topic has been published recently (Freeman & Steinmetz, 2011).

Eyelid closure is controlled by the facial motor nucleus (Pellegrini, Horn, & Evinger, 1995; Trigo, Gruart, & Delgado-Garcia, 1999). The UR in eyeblink conditioning can be elicited by the stimulation of the trigeminal nerve or trigeminal nucleus, which projects to the facial motor nucleus (Pellegrini et al., 1995; van Ham & Yeo, 1996a, 1996b). The facial motor nucleus also receives input from the red nucleus (Rosenfield, Dovydaitis, & Moore, 1985). The red nucleus in turn receives input from the contralateral cerebellar nuclei (Rosenfield et al., 1985).

As mentioned above, eyeblink conditioning depends upon the cerebellum. Damage to the anterior interpositus nucleus ipsilateral to the conditioned eye permanently blocks acquisition and retention of eyeblink conditioning (Clark, McCormick, Lavond, & Thompson, 1984; Lavond, Hembree, & Thompson, 1985; McCormick, Clark, Lavond, & Thompson, 1982; McCormick & Thompson, 1984a; Steinmetz, Lavond, Ivkovich, Logan, & Thompson, 1992; Steinmetz, Logue, & Steinmetz, 1992; Yeo,

Hardiman, & Glickstein, 1985); conditioning in the other eye is intact. Lesions of the contralateral red nucleus and facial motor nucleus have the same effect (Rosenfield & Moore, 1983, 1985). How then did investigators determine that the memory underlying eyeblink conditioning is stored in the cerebellum and not in downstream or upstream structures? The key to addressing this question was the use of reversible inactivation methods and careful behavioral analysis. Reversible inactivation has been used to silence the cerebellum, red nucleus, and motor nuclei during eyeblink conditioning. The logic of this approach is that a brain area may be storing the memory underlying eyeblink conditioning if the animal shows no evidence of acquisition during training and shows naive-like acquisition after inactivation is no longer administered (i.e., no savings). Conversely, inactivation of an area that is necessary for CR production but not memory storage will result in impaired CRs during training but complete savings of learning on the first session when inactivation is no longer administered. Using this approach, investigators found that reversible inactivation of the cerebellum completely blocked acquisition and savings (Freeman, Halverson, & Poremba, 2005; Krupa, Thompson, & Thompson, 1993). In contrast, inactivation of downstream structures including the superior cerebellar peduncle (output axons of the cerebellar nuclei), contralateral red nucleus, or motor nuclei impaired performance of the CR during acquisition, but the animals showed complete savings on the first training session without inactivation (Krupa & Thompson, 1995; Krupa et al., 1993; Krupa, Weng, & Thompson, 1996). Reversible inactivation or manipulations of synaptic function within the cerebellum after training sessions also impair memory consolidation (Attwell, Cooke, & Yeo, 2002; Cooke, Attwell, & Yeo, 2004). The findings of these reversible inactivation studies provide strong evidence that the memory underlying eyeblink conditioning is stored within the cerebellum and not in upstream or downstream structures (Fig. 4.2).

In vivo neurophysiology has provided essential information for understanding the relative contributions of the cerebellar cortex and nuclei to eyeblink conditioning and the cellular mechanisms underlying cerebellar memory formation. Extracellular recordings of neuronal activity showed that anterior interpositus neurons develop an increase in activity during training trials that correlates with the amplitude and time course of the CR both within trials and across training trials (Campolattaro, Kashef, Lee, & Freeman, 2011; Freeman & Nicholson, 2000; Halverson, Lee, & Freeman, 2010; McCormick & Thompson, 1984a, 1984b; Moore & Choi, 1997). The burst of interpositus neuronal activity within trials

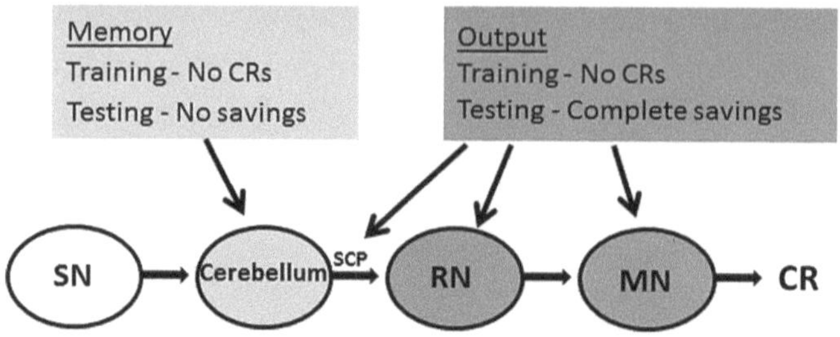

Figure 4.2 Diagram summarizing the effects of reversible inactivation on different parts of the neural circuitry necessary for eyeblink conditioning. Reversible inactivation of the cerebellum blocks conditional responses (CRs) during acquisition and results in no savings. In contrast, reversible inactivation of the red nucleus (RN), superior cerebellar peduncle (SCP), or motor nuclei (MN) that generate eyelid movement blocks CRs during acquisition but results in complete savings after inactivation is terminated. The findings indicate that the cerebellum is the primary site of memory storage and the SCP, RN, and MN are necessary for the output necessary to generate the eyelid closure CR.

precedes the CR by 30–50 ms, suggesting that it provides a premotor signal to activate downstream motor neurons. Further evidence for a causal role in anterior interpositus activity driving the CR comes from studies showing that electrical stimulation of the interpositus nucleus produces eyelid movement (Freeman & Nicholson, 2000; McCormick & Thompson, 1984a, 1984b). The *in vivo* neurophysiology thus indicates that neurons in the anterior interpositus nucleus undergo an increase in excitability during eyeblink conditioning, which is most likely due to increased synaptic input (Kleim et al., 2002; Ohyama, Nores, Medina, Riusech, & Mauk, 2006).

Purkinje cells in the cerebellar cortex also show learning-related changes during eyeblink conditioning. Purkinje cells that receive direct (i.e., short latency) US input from climbing fibers show decreases in simple spike activity during the CS (Green & Steinmetz, 2005; Hesslow & Ivarsson, 1994; Jirenhed, Bengtsson, & Hesslow, 2007; Jirenhed & Hesslow, 2011; Rasmussen, Jirenhed, & Hesslow, 2008). The decrease in simple spike activity releases the deep nuclei from inhibition momentarily during the CS, increasing interpositus nucleus activity and thereby increasing cerebellar output to the red nucleus and facial motor nucleus. The decrease in simple spike activity is thought to be the result of long-term depression at the parallel fiber-to-Purkinje cell synapses (Mauk & Donegan, 1997; Mauk, Garcia, Medina, & Steele, 1998; Raymond, Lisberger, & Mauk, 1996). The combination of synaptic enhancement at the mossy fiber-to-anterior interpositus nucleus synapses and synaptic depression at the parallel fiber-to-Purkinje cell synapses underlies the memory necessary for generating the CR (Mauk & Donegan, 1997; Medina, Nores, Ohyama, & Mauk, 2000).

3.1. CS pathway

Sensory input conveying the CS to the cerebellum originates in the various subcortical sensory nuclei. Animals that have been decorticated or have lesions of the primary sensory cortex show delay eyeblink conditioning but are impaired when the interval between the onset of the CS and onset of the US is greater than about 400 ms (Galvez, Weible, & Disterhoft, 2007; Steinmetz, Harmon, & Freeman, 2013). The auditory CS pathway includes sensory nucleus projections through the inferior colliculus to the medial auditory thalamus (Campolattaro, Halverson, & Freeman, 2007; Freeman, Halverson, & Hubbard, 2007; Halverson & Freeman, 2006, 2010a; Halverson et al., 2010; Halverson, Poremba, & Freeman, 2008). The medial auditory thalamus then projects to the lateral pontine nuclei (Campolattaro et al., 2007; Halverson & Freeman, 2010a), which then send their mossy fiber projection to the cerebellar cortex and nuclei. The visual CS pathway includes retinal projections to the ventral lateral geniculate and nucleus of the optic tract, which both project in parallel to the medial pontine nuclei (Halverson, Hubbard, & Freeman, 2009; Halverson & Freeman, 2010b; Steinmetz, Buss, & Freeman, 2013). The medial pontine nuclei then send a mossy fiber projection to the cerebellum. Stimulation of the mossy fiber pathway can be used as a CS in eyeblink conditioning when paired with a peripheral US such as periorbital stimulation (Freeman & Rabinak, 2004; Freeman, Rabinak, & Campolattaro, 2005; Steinmetz, Lavond, & Thompson, 1989; Steinmetz, Rosen, Chapman, Lavond, & Thompson, 1986; Steinmetz, Rosen, Woodruff-Pak, Lavond, & Thompson, 1986). The cerebellum sends feedback to the medial auditory thalamus and pontine nuclei during eyeblink conditioning.

3.2. US pathway

The US pathway includes a projection from the trigeminal nucleus to the inferior olive. The inferior olive then sends its climbing fiber axons into the cerebellar cortex and deep nuclei. Stimulation of the dorsal accessory division of the inferior olive can be used as a US in eyeblink conditioning when paired with a peripheral CS such as a tone or light (Mauk, Steinmetz, & Thompson, 1986; Steinmetz et al., 1989). The cerebellar nuclei provide inhibitory feedback to the inferior olive that dampens the climbing fiber response as learning develops (Andersson, Garwicz, & Hesslow, 1988; Bengtsson & Hesslow, 2006; Hesslow & Ivarsson, 1996; Sears & Steinmetz, 1991; Svensson, Bengtsson, & Hesslow, 2006). Thus, the same burst of anterior interpositus nucleus activity that drives the CR also inhibits the inferior olive. This inhibitory feedback is thought to limit plasticity mechanisms to the CS that predicts the US, thus precluding further conditioning to redundant or irrelevant stimuli (Kim, Krupa, & Thompson, 1998). Moreover, the stability of climbing fiber activity levels helps maintain plasticity within the cerebellum (Kenyon, Medina, & Mauk, 1998a, 1998b; Medina, Nores, & Mauk, 2002).

4. NEURAL MECHANISMS UNDERLYING THE ONTOGENY OF EYEBLINK CONDITIONING

Cerebellar cortical maturation continues through the period during which eyeblink conditioning emerges ontogenetically in rats. It is natural to assume then that granule cell neurogenesis and the continued development of synapses within the cerebellar cortex play a key role in the ontogeny of cerebellar learning. *In vivo* neuronal recordings from the anterior interpositus nucleus in developing rat pups indicate that the development of learning-related neuronal activity correlates with the ontogenetic emergence of eyeblink conditioning (Fig. 4.3) (Freeman & Nicholson, 2000). Younger pups that do not show robust eyeblink conditioning have fewer neurons that show learning-related activity. Among neurons that show learning-related activity, the magnitude of the activity burst is much lower in younger pups. The same type of developmental trend is seen in the cerebellar cortex, where fewer Purkinje cells show learning-related decreases in simple spike activity in younger pups (Nicholson & Freeman, 2004). These neurophysiological findings might suggest that the ontogeny of cerebellar learning is driven by developmental changes in plasticity mechanisms within the cerebellum. However, there are additional findings from these studies that suggest another developmental mechanism.

4.1. Development of the CS pathway

Cerebellar neuronal responses to the CS and US increase as a function of age. Thus, the developmental changes in learning-related activity could be driven by developmental changes in sensory inputs to the cerebellum rather than

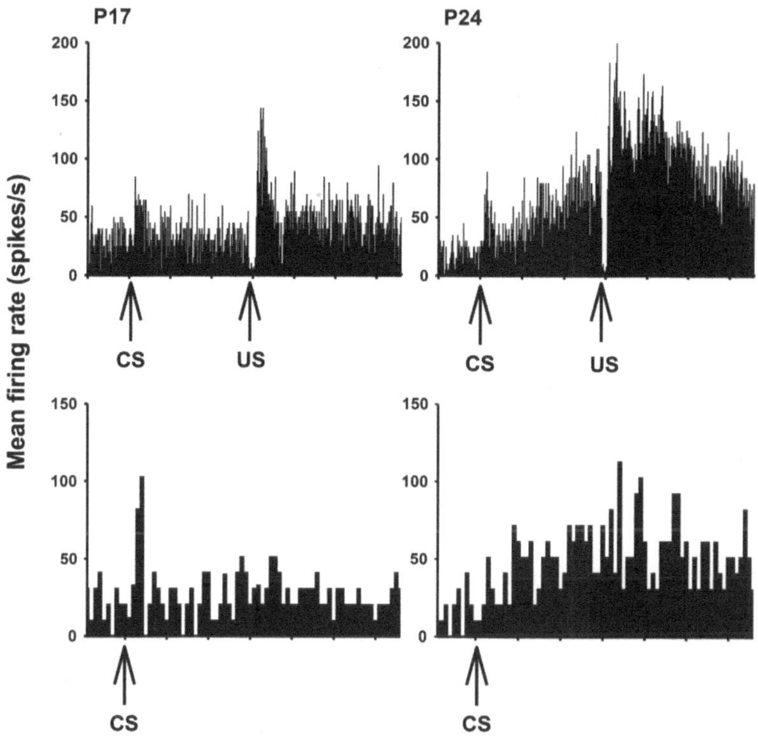

Figure 4.3 Development of learning-related activity in the anterior interpositus nucleus. Activity of single neurons recorded from rat pups given eyeblink conditioning on postnatal day (P)17 (left) or 24 (right). Neurons recorded at P24 show bursts of activity during the CS that correlate with the amplitude and time course of the conditional response (CR) and precede the CR within trials. Fewer neurons recorded at P17 show this learning-related activity.

developmental changes in cerebellar plasticity mechanisms. Neuronal recordings from the pontine nuclei are consistent with this hypothesis. Pontine neurons show a developmental increase in sensory responsiveness to auditory stimuli, suggesting that auditory input to the pontine nuclei might be maturing late (Freeman & Muckler, 2003). Developmental changes in sensory input to the pontine nuclei can be bypassed by using stimulation of the mossy fiber pathway as a CS in developing rats (Freeman, Halverson, & Poremba, 2005). This method produces robust eyeblink conditioning at ages that do not show learning with a tone or light CS (Fig. 4.4). Furthermore, mossy fiber stimulation produces eyeblink conditioning in 12-day-old rat pups that do not have fully open ear canals or eyes and have a relatively immature cerebellar cortex

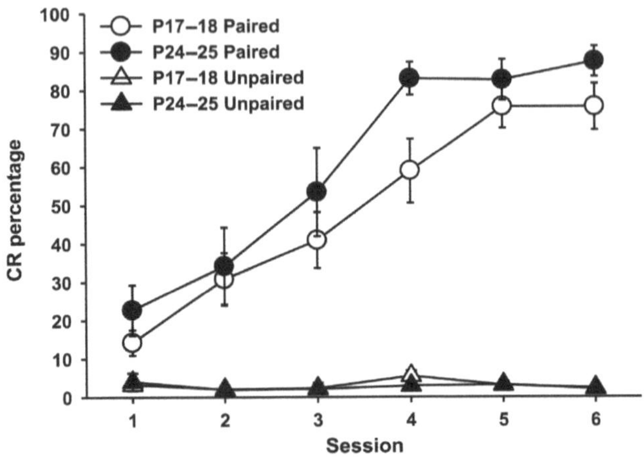

Figure 4.4 Early ontogenetic emergence of eyeblink conditioning using mossy fiber stimulation as the CS. Mean eyelid closure conditional response (CR) percentage for rat pups trained with pontine stimulation (left) or a tone (right) as the conditional stimulus (CS) on postnatal days (P)17–P18 (white symbols) or P24–P25 (black symbols). The pups were given either paired (circles) or unpaired (triangles) presentations of the CS and unconditional stimulus (US). Note that the group trained on P17–18 learns as well as the P24–25 group when given mossy fiber stimulation as the CS but not when a tone CS is used.

(Campolattaro & Freeman, 2008). The findings of these mossy fiber stimulation studies show that the cerebellum is capable of learning very early in ontogeny if given sufficiently strong mossy fiber input.

This stimulation method has also been applied to upstream auditory structures in an attempt to find a developmental bottleneck that may limit CS input to the cerebellum and thereby limit the induction of learning-related plasticity. Stimulation of the cochlear nucleus or medial auditory thalamus does not change the ontogenetic pattern seen with peripheral CSs (Freeman & Campolattaro, 2008; Freeman & Duffel, 2008) (Fig. 4.5). That is, stimulation of these auditory nuclei upstream from the pontine nuclei does not produce learning in younger rat pups as seen with mossy fiber stimulation. Since medial auditory thalamus stimulation does not produce learning in younger pups, the direct thalamopontine projection may be a developmental bottleneck limiting cerebellar plasticity. Neuronal recordings from the medial auditory thalamus indicate, however, that there are developmental changes upstream of the thalamus as well (Ng & Freeman, 2012). Developmental changes throughout the rat pup auditory system are currently under investigation using 2-DG imaging.

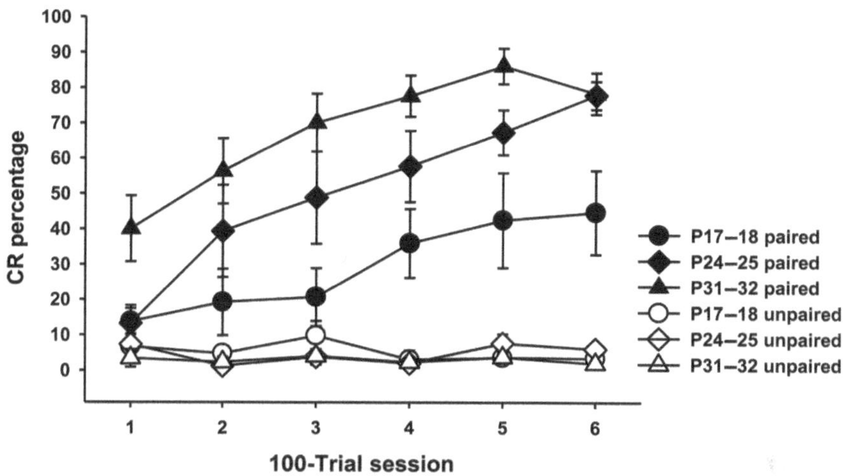

Figure 4.5 No early ontogenetic emergence of eyeblink conditioning with medial auditory thalamic stimulation as the CS. Mean eyelid conditional response (CR) percentage for rat pups trained with medial auditory thalamic stimulation as the conditional stimulus (CS) on P17–18 (circles), P24–25 (diamonds), or P31–32 (triangles). Pups were given either paired (black symbols) or unpaired (white symbols) presentations of the CS and an unconditional stimulus. Note that the CR percentage increases as a function of age in the paired groups.

4.2. Development of the US pathway

Developmental changes in CS input to the pontine nuclei clearly make a significant contribution to the ontogeny of cerebellar learning. But that is not the whole story. Although younger rat pups that do not learn with a peripheral CS learn with mossy fiber stimulation, the amplitude of the CRs is lower than in older pups. Thus, there are additional developmental mechanisms underlying the ontogeny of cerebellar learning. One such mechanism involves developmental changes in the US pathway. Neurons in the inferior olive and climbing fiber responses show a surprising developmental change: the neuronal response to the US is stronger in younger rat pups (Freeman & Nicholson, 2000; Nicholson & Freeman, 2003a, 2003b). This seemingly paradoxical finding can only be interpreted with reference to the neural simulation of cerebellar learning by Mauk and colleagues (Kenyon et al., 1998a, 1998b; Medina et al., 2002). In their model, inhibitory feedback regulates climbing fiber activity at an "equilibrium" level between training trials and sessions, which maintains synaptic plasticity established during training trials. Alterations of climbing fiber equilibrium will produce new synaptic plasticity that could deteriorate or saturate plasticity and result in impaired

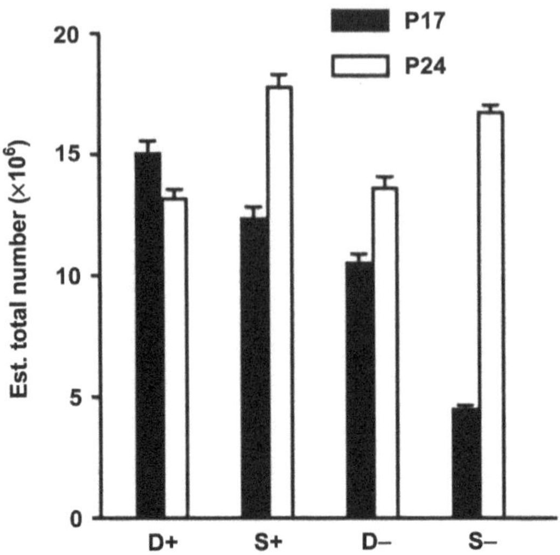

Figure 4.6 Development of inhibitory feedback from the cerebellum to the inferior olive. Mean total number of excitatory axodendritic (D+) and axospinous (S+) synapses; inhibitory axodendritic (D–) and axospinous (S–) synapses in the dorsal accessory inferior olive on P17 and P24. Note the age-related increase in D– and S– synapses. These data are from a quantitative electron microscopy study that used unbiased stereology.

learning. Neurophysiological and anatomical evidence indicates that there is a substantial developmental change in inhibitory feedback from the interpositus nucleus to the inferior olive that correlates with the developmental time course of eyeblink conditioning (Nicholson & Freeman, 2003a, 2003b) (Fig. 4.6). As a result of this developmental change in inferior olive inhibition, climbing fibers show more frequent and synchronized responses to the US in younger pups. Why would this impair cerebellar learning? The weaker inhibition of the inferior olive in younger pups produces a tonic increase in climbing fiber activity above the equilibrium level between trials and training sessions, resulting in the diminution of synaptic plasticity that was established during training trials. The weaker plasticity thereby produces weaker CRs in younger pups.

4.3. Development of forebrain modulation of cerebellar learning

Animals with surgical decerebration just rostral to the red nucleus can acquire eyeblink conditioning, but the rate of acquisition is slower than in intact animals. Thus, the forebrain is not necessary for eyeblink conditioning but facilitates acquisition. The amygdala is a forebrain area that has been shown to play a role in eyeblink conditioning. Lesions or inactivation of the amygdala impairs eyeblink conditioning and reflex facilitation (Blankenship, Huckfeldt, Steinmetz, & Steinmetz, 2005; Burhans, Smith-Bell, & Schreurs, 2008; Lee & Kim, 2004; Neufeld & Mintz, 2001). The amygdala is hypothesized to facilitate eyeblink conditioning through projections to the pontine nuclei (i.e., the CS pathway). A two-factor theory for amygdala modulation of eyeblink conditioning posits that the amygdala forms an initial fear-conditioning memory that then facilitates acquisition of a motor memory within the cerebellum (Lee & Kim, 2004; Neufeld & Mintz, 2001; Taub & Mintz, 2010). Amygdala-based fear conditioning can be observed earlier during ontogeny than eyeblink conditioning when freezing is measured but not when fear-potentiated startle is measured (Hunt, Richardson, & Campbell, 1994; Stanton, 2000). Amygdala modulation of the cerebellum may therefore play a role early in the ontogeny of eyeblink conditioning when freezing is seen or emerges later when fear-potentiated startle is seen. Reversible inactivation of the amygdala in rat pups produces an impairment in eyeblink conditioning as soon as CRs start to emerge developmentally (Ng & Freeman, 2012). Amygdala facilitation of cerebellar learning is therefore present as soon as the cerebellum–brainstem circuitry is mature enough to start producing eyeblink CRs. An important implication of this finding is that although the amygdala makes an important contribution to cerebellar learning, developmental changes in amygdala facilitation cannot account for the ontogeny of eyeblink conditioning. Contributions of septohippocampal theta to the ontogeny of cerebellar learning are currently under investigation.

5. CONCLUSIONS

The ontogeny of eyeblink conditioning depends on developmental changes in both the CS and US pathways (Fig. 4.7). Developmental changes in sensory inputs to the pontine nuclei from the medial auditory thalamus and upstream auditory nuclei limit the strength of CS input to the cerebellum via mossy fibers. The weaker mossy fiber input in younger animals results in weaker synaptic plasticity in the cerebellar cortex and anterior interpositus nucleus. The result of weaker plasticity is weaker learning. Developmental changes also occur in the inhibitory projection of the cerebellar nuclei to the inferior olive. This inhibitory feedback projection is necessary to prevent redundant learning and to maintain synaptic plasticity following conditioning trials. Younger animals lack this inhibitory feedback

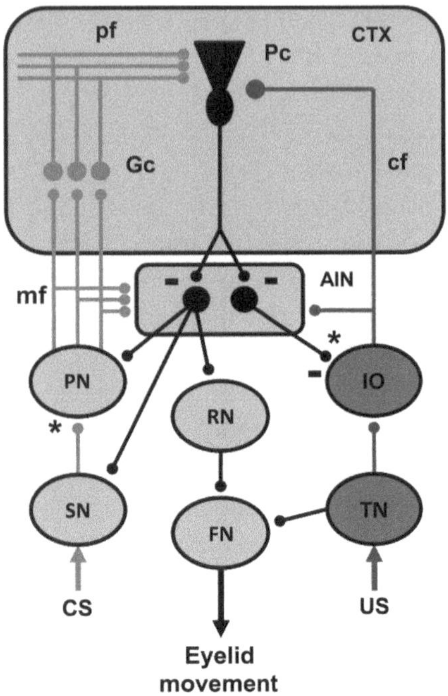

Figure 4.7 Diagram of the neural circuitry underlying eyeblink conditioning showing sites of developmental change. The cerebellar cortex (CTX) and anterior interpositus nucleus (AIN) receive input from the conditional stimulus (CS) through the pontine nucleus (PN) mossy fiber (mf) projection. The mf axons synapse on granule cells (Gc), which then send CS input to Purkinje cells (Pc) through parallel fibers (pf). The unconditional stimulus (US) reaches the cerebellum through climbing fibers (cf) originating in the inferior olive (IO). Inhibitory feedback regulation of the IO increases with age. Input to the pontine nuclei (PN) from subcortical sensory nuclei (SN) also increases with age, resulting in progressively stronger pf input to the Pc and mf input to the AIN. Weaker CS inputs combined with an unregulated IO lead to weaker synaptic plasticity in the cerebellum (AIN and Pc) and thereby weaker conditioning in younger rats. RN, red nucleus; FN, facial motor nucleus. Sites of developmental change in the circuitry are depicted by asterisks.

to the inferior olive and therefore have dysregulated climbing fiber activity, which results in weaker plasticity and thereby weaker learning. Amygdala modulation of the cerebellum is evident early in the development of eyeblink conditioning but does not play a critical role in the developmental processes underlying the ontogeny of eyeblink conditioning. Ongoing studies are further examining developmental changes in the CS pathway and in forebrain modulation of cerebellar learning.

REFERENCES

Altman, J. (1972a). Postnatal development of the cerebellar cortex in the rat. 3. Maturation of the components of the granular layer. *The Journal of Comparative Neurology*, *145*(4), 465–513. http://dx.doi.org/10.1002/cne.901450403.

Altman, J. (1972b). Postnatal development of the cerebellar cortex in the rat. I. The external germinal layer and the transitional molecular layer. *The Journal of Comparative Neurology*, *145*(3), 353–397. http://dx.doi.org/10.1002/cne.901450305.

Altman, J. (1972c). Postnatal development of the cerebellar cortex in the rat. II. Phases in the maturation of purkinje cells and of the molecular layer. *The Journal of Comparative Neurology*, *145*(4), 399–463. http://dx.doi.org/10.1002/cne.901450402.

Altman, J. (1975). Postnatal development of the cerebellar cortex in the rat. IV. Spatial organization of bipolar cells, parallel fibers and glial palisades. *The Journal of Comparative Neurology*, *163*(4), 427–447. http://dx.doi.org/10.1002/cne.901630404.

Altman, J., & Winfree, A. T. (1977). Postnatal development of the cerebellar cortex in the rat. V. Spatial organization of purkinje cell perikarya. *The Journal of Comparative Neurology*, *171*(1), 1–16. http://dx.doi.org/10.1002/cne.901710102.

Andersson, G., Garwicz, M., & Hesslow, G. (1988). Evidence for a GABA-mediated cerebellar inhibition of the inferior olive in the cat. *Experimental Brain Research. Experimentelle Hirnforschung. Experimentation Cerebrale*, *72*(3), 450–456.

Attwell, P. J., Cooke, S. F., & Yeo, C. H. (2002). Cerebellar function in consolidation of a motor memory. *Neuron*, *34*(6), 1011–1020.

Bengtsson, F., & Hesslow, G. (2006). Cerebellar control of the inferior olive. *Cerebellum (London, England)*, *5*(1), 7–14. http://dx.doi.org/10.1080/14734220500462757.

Blankenship, M. R., Huckfeldt, R., Steinmetz, J. J., & Steinmetz, J. E. (2005). The effects of amygdala lesions on hippocampal activity and classical eyeblink conditioning in rats. *Brain Research*, *1035*(2), 120–130. http://dx.doi.org/10.1016/j.brainres.2004.11.061.

Brown, K. L., & Woodruff-Pak, D. S. (2012). Eyeblink conditioning in the developing rabbit. *Developmental Psychobiology*, *54*(4), 423–432. http://dx.doi.org/10.1002/dev.20600.

Burhans, L. B., Smith-Bell, C., & Schreurs, B. G. (2008). Conditioning-specific reflex modification of the rabbit's nictitating membrane response and heart rate: Behavioral rules, neural substrates, and potential applications to posttraumatic stress disorder. *Behavioral Neuroscience*, *122*(6), 1191–1206. http://dx.doi.org/10.1037/a0013599.

Campolattaro, M. M., & Freeman, J. H. (2008). Eyeblink conditioning in 12-day-old rats using pontine stimulation as the conditioned stimulus. *Proceedings of the National Academy of Sciences of the United States of America*, *105*(23), 8120–8123. http://dx.doi.org/10.1073/pnas.0712006105.

Campolattaro, M. M., Halverson, H. E., & Freeman, J. H. (2007). Medial auditory thalamic stimulation as a conditioned stimulus for eyeblink conditioning in rats. *Learning & Memory (Cold Spring Harbor, N.Y.)*, *14*(3), 152–159. http://dx.doi.org/10.1101/lm.465507.

Campolattaro, M. M., Kashef, A., Lee, I., & Freeman, J. H. (2011). Neuronal correlates of cross-modal transfer in the cerebellum and pontine nuclei. *The Journal of Neuroscience*, *31*(11), 4051–4062. http://dx.doi.org/10.1523/JNEUROSCI.4142-10.2011.

Clark, G. A., McCormick, D. A., Lavond, D. G., & Thompson, R. F. (1984). Effects of lesions of cerebellar nuclei on conditioned behavioral and hippocampal neuronal responses. *Brain Research*, *291*(1), 125–136.

Cooke, S. F., Attwell, P. J., & Yeo, C. H. (2004). Temporal properties of cerebellar-dependent memory consolidation. *The Journal of Neuroscience*, *24*(12), 2934–2941. http://dx.doi.org/10.1523/JNEUROSCI.5505-03.2004.

Deaux, E. B., & Gormezano, I. (1963). Eyeball retraction: Classical conditioning and extinction in the albino rabbit. *Science (New York, N.Y.)*, *141*(3581), 630–631.

Freeman, J. H., & Campolattaro, M. M. (2008). Ontogenetic change in the auditory conditioned stimulus pathway for eyeblink conditioning. *Learning & Memory (Cold Spring Harbor, N.Y.)*, *15*(11), 823–828. http://dx.doi.org/10.1101/lm.1131208.

Freeman, J. H., & Duffel, J. W. (2008). Eyeblink conditioning using cochlear nucleus stimulation as a conditioned stimulus in developing rats. *Developmental Psychobiology*, *50*(7), 640–646. http://dx.doi.org/10.1002/dev.20331.

Freeman, J. H., Halverson, H. E., & Hubbard, E. M. (2007). Inferior colliculus lesions impair eyeblink conditioning in rats. *Learning & Memory (Cold Spring Harbor, N.Y.)*, *14*(12), 842–846. http://dx.doi.org/10.1101/lm.716107.

Freeman, J. H., Jr., Halverson, H. E., & Poremba, A. (2005). Differential effects of cerebellar inactivation on eyeblink conditioned excitation and inhibition. *The Journal of Neuroscience: The Official Journal of the Society for Neuroscience*, *25*(4), 889–895. http://dx.doi.org/10.1523/JNEUROSCI.4534-04.2005.

Freeman, J. H., Jr., & Muckler, A. S. (2003). Developmental changes in eyeblink conditioning and neuronal activity in the pontine nuclei. *Learning & Memory (Cold Spring Harbor, N.Y.)*, *10*(5), 337–345. http://dx.doi.org/10.1101/lm.63703.

Freeman, J. H., Jr., & Nicholson, D. A. (2000). Developmental changes in eye-blink conditioning and neuronal activity in the cerebellar interpositus nucleus. *The Journal of Neuroscience*, *20*(2), 813–819.

Freeman, J. H., Jr., & Rabinak, C. A. (2004). Eyeblink conditioning in rats using pontine stimulation as a conditioned stimulus. *Integrative Physiological and Behavioral Science*, *39*(3), 180–191.

Freeman, J. H., Jr., Rabinak, C. A., & Campolattaro, M. M. (2005). Pontine stimulation overcomes developmental limitations in the neural mechanisms of eyeblink conditioning. *Learning & Memory (Cold Spring Harbor, N.Y.)*, *12*(3), 255–259. http://dx.doi.org/10.1101/lm.91105.

Freeman, J. H., & Steinmetz, A. B. (2011). Neural circuitry and plasticity mechanisms underlying delay eyeblink conditioning. *Learning & Memory (Cold Spring Harbor, N.Y.)*, *18*(10), 666–677. http://dx.doi.org/10.1101/lm.2023011.

Galvez, R., Weible, A. P., & Disterhoft, J. F. (2007). Cortical barrel lesions impair whisker-CS trace eyeblink conditioning. *Learning & Memory (Cold Spring Harbor, N.Y.)*, *14*(1), 94–100. http://dx.doi.org/10.1101/lm.418407.

Gormezano, I., Schneiderman, N., Deaux, E., & Fuentes, I. (1962). Nictitating membrane: Classical conditioning and extinction in the albino rabbit. *Science (New York, N.Y.)*, *138*(3536), 33–34.

Green, J. T., & Steinmetz, J. E. (2005). Purkinje cell activity in the cerebellar anterior lobe after rabbit eyeblink conditioning. *Learning & Memory (Cold Spring Harbor, N.Y.)*, *12*(3), 260–269. http://dx.doi.org/10.1101/lm.89505.

Halverson, H. E., & Freeman, J. H. (2006). Medial auditory thalamic nuclei are necessary for eyeblink conditioning. *Behavioral Neuroscience*, *120*(4), 880–887. http://dx.doi.org/10.1037/0735-7044.120.4.880.

Halverson, H. E., & Freeman, J. H. (2010a). Medial auditory thalamic input to the lateral pontine nuclei is necessary for auditory eyeblink conditioning. *Neurobiology of Learning and Memory*, *93*(1), 92–98. http://dx.doi.org/10.1016/j.nlm.2009.08.008.

Halverson, H. E., & Freeman, J. H. (2010b). Ventral lateral geniculate input to the medial pons is necessary for visual eyeblink conditioning in rats. *Learning & Memory (Cold Spring Harbor, N.Y.)*, *17*(2), 80–85. http://dx.doi.org/10.1101/lm.1572710.

Halverson, H. E., Hubbard, E. M., & Freeman, J. H. (2009). Stimulation of the lateral geniculate, superior colliculus, or visual cortex is sufficient for eyeblink conditioning in rats. *Learning & Memory (Cold Spring Harbor, N.Y.)*, *16*(5), 300–307. http://dx.doi.org/10.1101/lm.1340909.

Halverson, H. E., Lee, I., & Freeman, J. H. (2010). Associative plasticity in the medial auditory thalamus and cerebellar interpositus nucleus during eyeblink conditioning. *The Journal of Neuroscience*, *30*(26), 8787–8796. http://dx.doi.org/10.1523/JNEUROSCI.0208-10.2010.

Halverson, H. E., Poremba, A., & Freeman, J. H. (2008). Medial auditory thalamus inactivation prevents acquisition and retention of eyeblink conditioning. *Learning & Memory (Cold Spring Harbor, N.Y.)*, *15*(7), 532–538. http://dx.doi.org/10.1101/lm.1002508.

Herbert, J. S., Eckerman, C. O., & Stanton, M. E. (2003). The ontogeny of human learning in delay, long-delay, and trace eyeblink conditioning. *Behavioral Neuroscience*, *117*(6), 1196–1210. http://dx.doi.org/10.1037/0735-7044.117.6.1196.

Hesslow, G., & Ivarsson, M. (1994). Suppression of cerebellar purkinje cells during conditioned responses in ferrets. *Neuroreport*, *5*(5), 649–652.

Hesslow, G., & Ivarsson, M. (1996). Inhibition of the inferior olive during conditioned responses in the decerebrate ferret. *Experimental Brain Research. Experimentelle Hirnforschung. Experimentation Cerebrale*, *110*(1), 36–46.

Hunt, P. S., Richardson, R., & Campbell, B. A. (1994). Delayed development of fear-potentiated startle in rats. *Behavioral Neuroscience*, *108*(1), 69–80.

Jirenhed, D. A., Bengtsson, F., & Hesslow, G. (2007). Acquisition, extinction, and reacquisition of a cerebellar cortical memory trace. *The Journal of Neuroscience*, *27*(10), 2493–2502. http://dx.doi.org/10.1523/JNEUROSCI.4202-06.2007.

Jirenhed, D. A., & Hesslow, G. (2011). Learning stimulus intervals—Adaptive timing of conditioned purkinje cell responses. *Cerebellum (London, England)*, *10*(3), 523–535. http://dx.doi.org/10.1007/s12311-011-0264-3.

Kenyon, G. T., Medina, J. F., & Mauk, M. D. (1998a). A mathematical model of the cerebellar-olivary system I: Self-regulating equilibrium of climbing fiber activity. *Journal of Computational Neuroscience*, *5*(1), 17–33.

Kenyon, G. T., Medina, J. F., & Mauk, M. D. (1998b). A mathematical model of the cerebellar-olivary system II: Motor adaptation through systematic disruption of climbing fiber equilibrium. *Journal of Computational Neuroscience*, *5*(1), 71–90.

Kim, J. J., Krupa, D. J., & Thompson, R. F. (1998). Inhibitory cerebello-olivary projections and blocking effect in classical conditioning. *Science (New York, N.Y.)*, *279*(5350), 570–573.

Kleim, J. A., Freeman, J. H., Jr., Bruneau, R., Nolan, B. C., Cooper, N. R., Zook, A., et al. (2002). Synapse formation is associated with memory storage in the cerebellum. *Proceedings of the National Academy of Sciences of the United States of America*, *99*(20), 13228–13231. http://dx.doi.org/10.1073/pnas.202483399.

Krupa, D. J., & Thompson, R. F. (1995). Inactivation of the superior cerebellar peduncle blocks expression but not acquisition of the rabbit's classically conditioned eye-blink response. *Proceedings of the National Academy of Sciences of the United States of America*, *92*(11), 5097–5101.

Krupa, D. J., Thompson, J. K., & Thompson, R. F. (1993). Localization of a memory trace in the mammalian brain. *Science (New York, N.Y.)*, *260*(5110), 989–991.

Krupa, D. J., Weng, J., & Thompson, R. F. (1996). Inactivation of brainstem motor nuclei blocks expression but not acquisition of the rabbit's classically conditioned eyeblink response. *Behavioral Neuroscience*, *110*(2), 219–227.

Lavond, D. G., Hembree, T. L., & Thompson, R. F. (1985). Effect of kainic acid lesions of the cerebellar interpositus nucleus on eyelid conditioning in the rabbit. *Brain Research*, *326*(1), 179–182.

Lee, T., & Kim, J. J. (2004). Differential effects of cerebellar, amygdalar, and hippocampal lesions on classical eyeblink conditioning in rats. *The Journal of Neuroscience*, *24*(13), 3242–3250. http://dx.doi.org/10.1523/JNEUROSCI.5382-03.2004.

Mauk, M. D., & Donegan, N. H. (1997). A model of pavlovian eyelid conditioning based on the synaptic organization of the cerebellum. *Learning & Memory (Cold Spring Harbor, N.Y.)*, *4*(1), 130–158.

Mauk, M. D., Garcia, K. S., Medina, J. F., & Steele, P. M. (1998). Does cerebellar LTD mediate motor learning? Toward a resolution without a smoking gun. *Neuron*, *20*(3), 359–362.

Mauk, M. D., Steinmetz, J. E., & Thompson, R. F. (1986). Classical conditioning using stimulation of the inferior olive as the unconditioned stimulus. *Proceedings of the National Academy of Sciences of the United States of America*, *83*(14), 5349–5353.

McCormick, D. A., Clark, G. A., Lavond, D. G., & Thompson, R. F. (1982). Initial localization of the memory trace for a basic form of learning. *Proceedings of the National Academy of Sciences of the United States of America*, *79*(8), 2731–2735.

McCormick, D. A., & Thompson, R. F. (1984a). Neuronal responses of the rabbit cerebellum during acquisition and performance of a classically conditioned nictitating membrane-eyelid response. *The Journal of Neuroscience*, *4*(11), 2811–2822.

McCormick, D. A., & Thompson, R. F. (1984b). Cerebellum: Essential involvement in the classically conditioned eyelid response. *Science (New York, N.Y.)*, *223*(4633), 296–299.

Medina, J. F., Nores, W. L., & Mauk, M. D. (2002). Inhibition of climbing fibres is a signal for the extinction of conditioned eyelid responses. *Nature*, *416*(6878), 330–333. http://dx.doi.org/10.1038/416330a.

Medina, J. F., Nores, W. L., Ohyama, T., & Mauk, M. D. (2000). Mechanisms of cerebellar learning suggested by eyelid conditioning. *Current Opinion in Neurobiology*, *10*(6), 717–724.

Moore, J. W., & Choi, J. S. (1997). Conditioned response timing and integration in the cerebellum. *Learning & Memory (Cold Spring Harbor, N.Y.)*, *4*(1), 116–129.

Neufeld, M., & Mintz, M. (2001). Involvement of the amygdala in classical conditioning of eyeblink response in the rat. *Brain Research*, *889*(1–2), 112–117.

Ng, K. H., & Freeman, J. H. (2012). Developmental changes in medial auditory thalamic contributions to associative motor learning. *The Journal of Neuroscience*, *32*(20), 6841–6850. http://dx.doi.org/10.1523/JNEUROSCI.0284-12.2012.

Nicholson, D. A., & Freeman, J. H., Jr. (2003a). Addition of inhibition in the olivocerebellar system and the ontogeny of a motor memory. *Nature Neuroscience*, *6*(5), 532–537. http://dx.doi.org/10.1038/nn1042.

Nicholson, D. A., & Freeman, J. H., Jr. (2003b). Developmental changes in evoked purkinje cell complex spike responses. *Journal of Neurophysiology*, *90*(4), 2349–2357. http://dx.doi.org/10.1152/jn.00481.2003.

Nicholson, D. A., & Freeman, J. H., Jr. (2004). Developmental changes in eyeblink conditioning and simple spike activity in the cerebellar cortex. *Developmental Psychobiology*, *44*(1), 45–57. http://dx.doi.org/10.1002/dev.10149.

Ohyama, T., Nores, W. L., Medina, J. F., Riusech, F. A., & Mauk, M. D. (2006). Learning-induced plasticity in deep cerebellar nucleus. *The Journal of Neuroscience*, *26*(49), 12656–12663. http://dx.doi.org/10.1523/JNEUROSCI.4023-06.2006.

Pellegrini, J. J., Horn, A. K., & Evinger, C. (1995). The trigeminally evoked blink reflex. I. Neuronal circuits. *Experimental Brain Research. Experimentelle Hirnforschung. Experimentation Cerebrale*, *107*(2), 166–180.

Rasmussen, A., Jirenhed, D. A., & Hesslow, G. (2008). Simple and complex spike firing patterns in purkinje cells during classical conditioning. *Cerebellum (London, England)*, *7*(4), 563–566. http://dx.doi.org/10.1007/s12311-008-0068-2.

Raymond, J. L., Lisberger, S. G., & Mauk, M. D. (1996). The cerebellum: A neuronal learning machine? *Science (New York, N.Y.)*, *272*(5265), 1126–1131.

Rosenfield, M. E., Dovydaitis, A., & Moore, J. W. (1985). Brachium conjuntivum and rubrobulbar tract: Brain stem projections of red nucleus essential for the conditioned nictitating membrane response. *Physiology & Behavior*, *34*(5), 751–759.

Rosenfield, M. E., & Moore, J. W. (1983). Red nucleus lesions disrupt the classically conditioned nictitating membrane response in rabbits. *Behavioural Brain Research*, *10*(2–3), 393–398.
Rosenfield, M. E., & Moore, J. W. (1985). Red nucleus lesions impair acquisition of the classically conditioned nictitating membrane response but not eye-to-eye savings or unconditioned response amplitude. *Behavioural Brain Research*, *17*(1), 77–81.
Schneiderman, N., Fuentes, I., & Gormezano, I. (1962). Acquisition and extinction of the classically conditioned eyelid response in the albino rabbit. *Science (New York, N.Y.)*, *136*(3516), 650–652.
Sears, L. L., & Steinmetz, J. E. (1991). Dorsal accessory inferior olive activity diminishes during acquisition of the rabbit classically conditioned eyelid response. *Brain Research*, *545*(1–2), 114–122.
Stanton, M. E. (2000). Multiple memory systems, development and conditioning. *Behavioural Brain Research*, *110*(1–2), 25–37.
Stanton, M. E., Freeman, J. H., Jr., & Skelton, R. W. (1992). Eyeblink conditioning in the developing rat. *Behavioral Neuroscience*, *106*(4), 657–665.
Steinmetz, A. B., Buss, E. W., & Freeman, J. H. (2013). Inactivation of the ventral lateral geniculate and nucleus of the optic tract impairs retention of visual eyeblink conditioning. *Behavioral Neuroscience*, *127*(5), 690–693. http://dx.doi.org/10.1037/a0033729.
Steinmetz, A. B., Harmon, T. C., & Freeman, J. H. (2013). Visual cortical contributions to associative cerebellar learning. *Neurobiology of Learning and Memory*, *104*, 103–109. http://dx.doi.org/10.1016/j.nlm.2013.06.005.
Steinmetz, J. E., Lavond, D. G., Ivkovich, D., Logan, C. G., & Thompson, R. F. (1992). Disruption of classical eyelid conditioning after cerebellar lesions: Damage to a memory trace system or a simple performance deficit? *The Journal of Neuroscience*, *12*(11), 4403–4426.
Steinmetz, J. E., Lavond, D. G., & Thompson, R. F. (1989). Classical conditioning in rabbits using pontine nucleus stimulation as a conditioned stimulus and inferior olive stimulation as an unconditioned stimulus. *Synapse (New York)*, *3*(3), 225–233. http://dx.doi.org/10.1002/syn.890030308.
Steinmetz, J. E., Logue, S. F., & Steinmetz, S. S. (1992). Rabbit classically conditioned eyelid responses do not reappear after interpositus nucleus lesion and extensive post-lesion training. *Behavioural Brain Research*, *51*(1), 103–114.
Steinmetz, J. E., Rosen, D. J., Chapman, P. F., Lavond, D. G., & Thompson, R. F. (1986). Classical conditioning of the rabbit eyelid response with a mossy-fiber stimulation CS: I. Pontine nuclei and middle cerebellar peduncle stimulation. *Behavioral Neuroscience*, *100*(6), 878–887.
Steinmetz, J. E., Rosen, D. J., Woodruff-Pak, D. S., Lavond, D. G., & Thompson, R. F. (1986). Rapid transfer of training occurs when direct mossy fiber stimulation is used as a conditioned stimulus for classical eyelid conditioning. *Neuroscience Research*, *3*(6), 606–616.
Svensson, P., Bengtsson, F., & Hesslow, G. (2006). Cerebellar inhibition of inferior olivary transmission in the decerebrate ferret. *Experimental Brain Research. Experimentelle Hirnforschung. Experimentation Cerebrale*, *168*(1–2), 241–253. http://dx.doi.org/10.1007/s00221-005-0086-y.
Taub, A. H., & Mintz, M. (2010). Amygdala conditioning modulates sensory input to the cerebellum. *Neurobiology of Learning and Memory*, *94*(4), 521–529. http://dx.doi.org/10.1016/j.nlm.2010.09.004.
Trigo, J. A., Gruart, A., & Delgado-Garcia, J. M. (1999). Discharge profiles of abducens, accessory abducens, and orbicularis oculi motoneurons during reflex and conditioned blinks in alert cats. *Journal of Neurophysiology*, *81*(4), 1666–1684.

van Ham, J. J., & Yeo, C. H. (1996a). The central distribution of primary afferents from the external eyelids, conjunctiva, and cornea in the rabbit, studied using WGA-HRP and B-HRP as transganglionic tracers. *Experimental Neurology, 142*(2), 217–225. http://dx.doi.org/10.1006/exnr.1996.0193.

van Ham, J. J., & Yeo, C. H. (1996b). Trigeminal inputs to eyeblink motoneurons in the rabbit. *Experimental Neurology, 142*(2), 244–257. http://dx.doi.org/10.1006/exnr.1996.0195.

Yeo, C. H., Hardiman, M. J., & Glickstein, M. (1985). Classical conditioning of the nictitating membrane response of the rabbit. I. Lesions of the cerebellar nuclei. *Experimental Brain Research. Experimentelle Hirnforschung. Experimentation Cerebrale, 60*(1), 87–98.

INDEX

Note: Page numbers followed by "*f*" indicate figures.

CONTENTS OF RECENT VOLUMES

Volume 37

Volume 38

Volume 42

Volume 43

Volume 44

Volume 45

Volume 46

Volume 47

Processing of Retinal Tissue for *in Situ* Hybridization
Frank Müller

Processing the Spinal Cord for *in Situ* Hybridization with Radiolabelled Oligonucleotides
A. Berthele and T. R. Tölle

Processing Human Brain Tissue for *in Situ* Hybridization with Radiolabelled Oligonucleotides
Louise F. B. Nicholson

In Situ Hybridization of Astrocytes and Neurons Cultured *in Vitro*
L. A. Arizza-McNaughton, C. De Felipe, and S. P. Hunt

In Situ Hybridization on Organotypic Slice Cultures
A. Gerfin-Moser and H. Monyer

Quantitative Analysis of *in Situ* Hybridization Histochemistry
Andrew L. Gundlach and Ross D. O'Shea

Part II: Nonradioactive *in Situ* hybridization

Nonradioactive *in Situ* Hybridization Using Alkaline Phosphatase-Labelled Oligonucleotides
S. J. Augood, E. M. McGowan, B. R. Finsen, B. Heppelmann, and P. C. Emson

Combining Nonradioactive *in Situ* Hybridization with Immunohistological and Anatomical Techniques
Petra Wahle

Nonradioactive *in Situ* Hybridization: Simplified Procedures for Use in Whole Mounts of Mouse and Chick Embryos
Linda Ariza-McNaughton and Robb Krumlauf

INDEX

Volume 48

Assembly and Intracellular Trafficking of $GABA_A$ Receptors Eugene
Barnes

Subcellular Localization and Regulation of $GABA_A$ Receptors and Associated Proteins Bernhard Lüscher and Jean-Marc Fritschy D_1 Dopamine Receptors
Richard Mailman

Molecular Modeling of Ligand-Gated Ion Channels: Progress and Challenges
Ed Bertaccini and James R. Trudel

Alzheimer's Disease: Its Diagnosis and Pathogenesis
Jillian J. Kril and Glenda M. Halliday

DNA Arrays and Functional Genomics in Neurobiology
Christelle Thibault, Long Wang, Li Zhang, and Michael F. Miles

INDEX

Volume 49

What Is West Syndrome?
Olivier Dulac, Christine Soufflet, Catherine Chiron, and Anna Kaminski

The Relationship between encephalopathy and Abnormal Neuronal Activity in the Developing Brain
Frances E. Jensen

Hypotheses from Functional Neuroimaging Studies
Csaba Juhász, Harry T. Chugani, Ouo Muzik, and Diane C. Chugani

Infantile Spasms: Unique Sydrome or General Age-Dependent Manifestation of a Diffuse Encephalopathy?
M. A. Koehn and M. Duchowny

Histopathology of Brain Tissue from Patients with Infantile Spasms
Harry V. Vinters

Generators of Ictal and Interictal Electroencephalograms Associated with Infantile Spasms: Intracellular Studies of Cortical and Thalamic Neurons
M. Steriade and I. Timofeev

Cortical and Subcortical Generators of Normal and Abnormal Rhythmicity
David A. McCormick

Role of Subcortical Structures in the Pathogenesis of Infantile Spasms: What Are Possible Subcortical Mediators?
F. A. Lado and S. L. Moshé

What Must We Know to Develop Better Therapies?
Jean Aicardi

Volume 50

Volume 51

Volume 52

Volume 53

Volume 54

Volume 55

Volume 56

Volume 57

Volume 58

Volume 59

Volume 60

Volume 61

Volume 62

Volume 63

Volume 64

Volume 65

Volume 66

Volume 67

Volume 68

Volume 69

Volume 70

Volume 71

Volume 72

Volume 73

Volume 74

Volume 75

Volume 76

Volume 77

Volume 78

Volume 79

Volume 80

Volume 81

Volume 82

Volume 83

Volume 84

Volume 85

Volume 86

Volume 87

Volume 90

Volume 91

Volume 92

Volume 93

Volume 94

Volume 95

Volume 96

Volume 97

Volume 98

Volume 99

Volume 100

Volume 101

Volume 102

Volume 103

Volume 104

Volume 105

Volume 106

Volume 107

Volume 108

Volume 109

Volume 110

Volume 111

Volume 112

Volume 113

Volume 114

Volume 115

Volume 116

FSC
www.fsc.org
MIX
Papier | Fördert gute Waldnutzung
FSC® C083411

Kolibri 360 GmbH
Ferdinand-Jühlke-Straße 7
99095 Erfurt, Deutschland
produktsicherheit@kolibri360.de